MINISTRY OF AGRICULTURE, FISHERIES AND FOOD
AGRICULTURAL DEVELOPMENT AND ADVISORY SERVICE
SLOUGH LABORATORY

Reference Book 437

INSECT TRAVELLERS

VOLUME II

by

Audrey D Aitken

A survey of insects,
excluding beetles, recorded from imported cargoes, by the
Insect Inspectorate of the Ministry of Agriculture, Fisheries and Food
during the period 1957 to 1977.

LONDON: HER MAJESTY'S STATIONERY OFFICE

Foreword

Insect Travellers Volume I appeared in 1975. It provided for the first time a comprehensive survey of the beetles recorded from imported cargoes and presented in an accessible form a wealth of information, much of which had been derived over more than a decade from the activities of the Insect Infestation Inspectorate of the Ministry of Agriculture, Fisheries and Food. It has since found its way on to the reference shelves of most storage entomologists active in the United Kingdom and many overseas. However, it dealt only with the Coleoptera. Readers of Volume I will have realised that Mrs Aitken has been continuing her work and that a second volume dealing with the other orders has been in preparation. Volume II now completes the picture first revealed with the publication of volume I.

Although the Coleoptera contribute more species than any other order to the list of insects recorded from imported cargoes, the Lepidoptera come next in order of frequency and include a number of species of considerable commercial inportance as pests of food, animal feedingstuffs, or other stored commodities. Some of the species of Hemiptera and Hymenoptera found in stored products have a commercial importance because of the effects they may have as predators or parasites of pests.

Readers who are interested in the relationships between insects and public health will no doubt find food for thought in the Appendix which lists many species of exotic cockroaches and ants as stray introductions found among stored products. Some of these species undoubtedly have the potential to establish themselves here as public health pests.

Pests tend to become cosmopolitan; this is perhaps especially true of pests of stored products. Insect Travellers reveals not only how wide-ranging the geographical distrubutions of some insect species have become, but also throws light on the means by which their dispersal has been achieved.

Information of the kind presented in this book is now collected only selectively and accrues much more slowly, partly as a result of the advent of containerisation as discussed by Mrs Aitken in her Introduction and partly because of changes of emphasis in the objective of the Ministry staff involved. The Insect Infestation Inspectorate no longer exists. The efforts of its successor, which has recently been re-formed into the Wildlife and Storage Biology Discipline of the Agricultural Science Service of the Ministry's Agricultural Development and Advisory Service, are directed much more in support of agricultural production. For these reasons Mrs Aitken's work is unlikely to be superseded and, now that it is complete, it is likely to remain the standard reference on the subject for some time.

F. A. Hunter

Acknowledgements

A large part of the work on this second volume was accomplished after I had left the Ministry of Agriculture's employment and I am greatly indebted to my former colleagues within the Ministry for their continued co-operation.

In particular I wish to thank Mr Richard Adams, who is currently in charge of insect identification and reference collection at the Ministry's Slough Laboratory, and also Mrs Margaret Stratford, lately librarian, and her staff for their tireless efforts in keeping me supplied with literature and information. Special thanks are due to Mr Eric Hurlock for the wealth of information contained in the analyses of ship inspections, prepared under his direction, and for his invaluable assistance in the final preparation of my manuscript for publication.

Contents

Introduction

This is the second of two volumes devoted to the insects that travel the world in cargoes of foodstuffs, in freight containers and in the holds of cargo ships. In volume I (Aitken, 1975) I dealt solely with the species of Coleoptera. In this second volume I list and comment on the recorded species of other orders of insects, and include 276 species from 11 orders.

In the introduction to volume I, I described the aims of this survey as the provision of a comprehensive list of the insects recorded from imports by the Inspectorate of the Ministry of Agriculture, Fisheries and Food and of the Department of Agriculture and Fisheries for Scotland. This aim is fulfilled by the completion of this second volume.

The sources of records

The records listed here derive from the written reports of the Insect Inspectorate of the Ministry of Agriculture and of the Department of Agriculture for Scotland, whose work it was to examine food cargoes such as grain and animal feedingstuffs and some other infestible commodities such as animal products that are brought into the country in cargo ships and containers. I have not included records of insects on plant and vegetable cargoes obtained by the Plant Health and Seeds Inspectorate.

Specimens of the commonest species were, in the main, identified by the inspectors who found them. The less common and the rare species were identified by entomologists at the Ministry's laboratories at Tolworth and Slough, or by various specialists working at the British Museum or elsewhere. Much useful comment was contributed by these specialists and, where relevant, I have quoted this material as 'identifiers' remarks'. In many cases, the specimens themselves are preserved in the Ministry's reference collection which is now housed at the Slough Laboratory.

Over the years covered by this survey, from the earliest ship inspections in the 1940s until 1977, there were many thousands of inspections and written reports. The insect records from all written reports since 1957 have been analysed and summarised annually as tabulated analyses, from the different countries of the world and on different types of cargo. It is these yearly analyses that have formed the basis of my tables of importations recorded against the more common species.

The effect of containerisation

The tables in volume I analysed the infestations recorded during the years 1957 to 1969. For volume II, I have extended this period for a further eight years, to 1977. These later years saw a progressive change in trading practice, away from the stowage of cargoes in the holds of ships, towards stowage in freight containers. This has led to a gradual decline in the traditional ports, where goods were unloaded piecemeal into lighters or on to road transport, and where cargoes were easily accessible for inspection and, if necessary, for insecticidal treatment, before distribution to their final destination. Instead, loaded containers are distributed unopened from large depots to their destinations, which may be on farms, mills or factories, often far inland. The opportunity for inspection and treatment at source is lost and the countrywide spread of infestation is facilitated. The old style of ship and wharf inspection had virtually ceased by the late 1970s to be replaced by increased vigilance inland. Records and analyses made after 1977, therefore, are not comparable with those made before.

The presentation of records

In volume I, pest species, predators and accidental strays were all listed together, each one accorded some comment on its world distribution and, when known, on its habits and biology. In this second volume, however, the main list is reserved for those species that are directly related, as pests, to the cargoes on which they were found or, as parasites or predators, to the pest population. Notable amongst the pests are the Lepidoptera, in particular the species of *Ephestia*, *Plodia interpunctella* and *Corcyra cephalonica*. Important parasites and predators are found in the Hymenoptera and Hemiptera.

Accidental strays are listed in a separate appendix. This does not necessarily mean that they are of no economic importance, only that they are not primarily associated with imported food cargoes. Some of them can, in certain circumstances, become serious environmental pests. Cockroaches, some flies and ants such as *Monomorium pharaonis* (L.) are known pests, even though in this survey they are relegated to a secondary list.

Table 1

The number of cargoes in different commodity groups from different geographic regions examined by the Ministry's Inspectorate during the 21 years 1957 to 1977.

	Nearctic	Neotropical	Ethiopian	Northern Europe	Mediterranean	Oriental	Australasian	Totals
Grain	12019	2470	1369	6839	2066	2025	4575	31363
Grain products	3699	1145	2357	619	643	2240	2540	13243
Oilseeds	700	567	7321	275	591	1292	65	10811
Copra	0	1	137	0	0	250	128	516
Oilcake	427	2251	5354	474	1114	7611	115	17346
Cocoa beans	0	957	2062	0	0	186	483	3688
Dried fruit	2191	272	1104	167	7045	344	1653	12776
Nuts	530	592	368	807	5908	1635	122	9962
Pulse and pulse products	4300	361	2776	663	1230	851	852	11033
Carobs and carob meal	0	0	0	0	2103	7	0	2110
Spices	20	628	1498	82	253	3655	60	6196
Illipenuts	0	0	0	0	0	150	0	150
Sago, sago flour	0	0	5	0	2	2118	0	2125
Animal products	1022	2141	2683	1065	514	2682	2039	12146
Others	1501	1815	5684	736	1223	6465	865	18289
Totals	26409	13200	32718	11727	22692	31511	13497	151754

Although I list no records later than 1977, every attempt has been made to use the latest nomenclature. Common names have been taken from Seymour (1979).

The presentation of records in tabulated form is exactly the same as in volume II with the number of occurrences cross-referenced to the type of foodstuff and the zoogeographic region. The seven zoogeographic regions are based on the 'natural regions' of Jeannel (1960). In Table 1 I analyse the number of the different types of cargo from the different geographic regions for the 21-year-period 1957 to 1977.

List of Recorded Insects

Thysanura

These are very primitive insects, wholly wingless and showing the basic insect structure in its simplest form. Their bodies are covered in scales and the abdomen terminates in three cerci, from which the common name, bristletail, derives.

Lepisma saccharina (L.) – – Silverfish
> *Distribution:* cosmopolitan. Although it has been considered as originating in warm areas, it is able to survive rigorous winter conditions in Britain and is well adapted to living in unheated buildings (Solomon and Adamson, 1955).

Interceptions: (see Table 2)

Table 2

Records of Lepisma saccharina *(L.)*
Total number of records = 59

Commodity group	Number of records	Geographic region	Number of records
Grain	3	Nearctic	2
Grain products	4	Neotropical	2
Oilcake	26	Ethiopian	8
Dried fruit	5	Northern Europe	0
Animal products	8	Mediterranean	5
Others	13	Oriental	40
		Australasian	2

India, which accounted for 28 of these records, was by far the most frequent source of this species.

> *Biology:* these are nocturnal insects which often occur in damp places and which require a high humidity to thrive. In houses, they do little actual damage but are a nuisance. They feed on carbohydrate substances such as starch used as wallpaper paste and are probably able to digest the cellulose in damp wallpaper. They are also recorded on fabrics and probably supplement their diet with protein from dead insects, size and glues from book bindings etc. (Hickin, 1974). Woodroffe (1953) records them, sometimes in large numbers, in feral pigeon nests, where it is surprising to find this supposedly starch feeder thriving on solid pigeon guano.

Thermobia domestica (Packard) – – firebrat, bristletail
> *Distribution:* throughout the warm regions of the world. Occurs in temperate regions only in heated surroundings.

Interceptions: (see Table 3)

Table 3

Records of Thermobia domestica *(Packard)*

Total number of records = 141

Commodity group	Number of records	Geographic region	Number of records
Grain	15	Nearctic	4
Grain products	21	Neotropical	11
Oilcake	58	Ethiopian	13
Dried fruit	13	Northern Europe	1
Animal products	12	Mediterranean	13
Others	22	Oriental	99
		Australasian	0

India and Pakistan together accounted for 80 of these records, and were the most frequent sources of this species.

Biology: Thermobia domestica feeds on similar food to *Lepisma saccharina* but, unlike that species, a high humidity is not necessary for its welfare. It requires a high temperature, however, and in Britain and other temperate countries it occurs only in very warm, even hot, surroundings such as bakeries, restaurants, kitchens or factories where steam is used. (Hicken, 1974).

Orthoptera — Saltatoria

The Orthoptera-Saltatoria include crickets, grasshoppers and locusts, the best known feature of which are the greatly enlarged back legs, which are used for hopping. The adult males stridulate by rubbing two finely ridged areas of cuticle together — the hind femur rubbed on the wing in the case of grasshoppers and locusts, or the wings rubbed together in the crickets (Tettigoniidae and Gryllidae).

Gryllidae

These are the crickets, many species of which have been shown to be omnivorous, able to feed on insects or other small arthropods, substances of animal origin as well as on plant tissue. Some species (e.g *Acheta domesticus, Gryllodes sigillatus*) have become domiciliary species, living as scavengers in dwellings and other buildings.

Acheta domesticus (L.) – – house cricket
  (= *Gryllus domesticus* L.)
  (= *Gryllulus domesticus* L.)
  Distribution: Cosmopolitan. This species probably originated in very dry areas of North Africa and South West Asia, where it occurs out of doors. Cornes (1973) records it in Nigeria. In Britain and other temperate countries it can only survive in warm conditions, heated buildings, fermenting refuse etc.

Interceptions:
   Turkey: lentils
   Nigeria: sheanuts
   West Africa: cocoa beans
   Iran: wheat bran
   India: cottonseed cake, bones

Biology: this is the house cricket. It is a nocturnal insect and occurs in Britain as a scavenger in heated buildings such as bakeries, restaurants, factories, kitchens and near boilers, as well as in coal mines. Out of doors in this country it is often found on rubbish dumps where fermenting and decaying refuse provides the necessary warmth for it to survive the winter. In warm summers, crickets migrating from rubbish dumps may invade nearby houses, where they can cause considerable annoyance, not least by the incessant chirping of the males.

The house cricket is omnivorous, will feed on almost any foodstuffs and has been reported damaging textiles, including artificial fibres, leather and wood. It will also do considerable damage to plant seedlings out of doors in warm regions and in greenhouses. It is also known to be a predator; Pimentel and Cranston (1960) discuss the species as a predator of house fly pupae. Breeding occurs within the temperature range 25 to 35°C. Nymphal development may be completed, albeit slowly, at 40°C although this temperature is unfavourably high. Nymphal development is quickest (23 to 33 days) at 35°C but may take over 200 days at lower temperatures. (Busvine, 1955; Ghouri and McFarlane, 1958; Ragge, 1965; Hickin, 1974).

Gryllodes sigillatus (Walker)

Distribution: widespread in tropical and subtropical regions.

Interceptions:
   East Africa: coconuts and coconut shell, sawn timber and cottonseed cake
   Nigeria: bones, cocoa beans
   Cameroons: hold of ship after cargo of coffee and tobacco
   Singapore: timber

Biology: this species has been recorded as a common household pest in India, where it causes damage to textiles. The eggs are laid in soft earth, under damp paper or in food (Khan, 1954). Gourhi and McFarlane (1958) found that nymphal development ranged from 75 to 155 days at 23°C and 27 to 40 days at 35°C (the optimum temperature).

Gryllulus domesticus L., see *Acheta domesticus* L.

Gryllus bimaculatus (De Geer)

Distribution: widespread in tropical and subtropical countries of Asia, Africa and Europe.

Interceptions:
   Russia: barley
   Spain or Canary Islands: canned peach pulp
   Tunisia: grass pulp
   East Africa: coconuts and coconut shell, sawn timber, cottonseed cake

South Africa: tinned peaches
India: ginger root

Biology: this species has been recorded damaging seedlings of cotton, rice and tobacco, sugar cane, pea pods and maize cobs. Although the species is not often mentioned in literature, when it is, it is usually referred to as an important pest. It seems, therefore, that outbreaks are spasmodic. Occasional large populations have been reported in Formosa and the Philippines, Italy and Israel. In Israel it has been reported in such numbers as to resemble a swarm of locust hoppers, attacking soft maize cobs, but this degree of infestation is exceptional. Development and breeding is possible within a temperature range of 19 to 34°C. At 34°C total development takes 22 to 44 days while at 22°C it takes over four months. In Israel three generations a year may be produced (Rivnay and Ziv, 1963).

Gryllus domesticus L., see *Acheta domesticus* L.

Dermaptera

These are earwigs, readily recognised by the modification of the cerci of the adults into unjointed horny forceps. Some are wingless but others are capable of flight and bear membranous hind wings that can be folded beneath the forewings, which are modified to form very short leathery tegmina. Earwigs are unusual amongst insects in that the female exhibits maternal care of the eggs and newly hatched nymphs, digging a special brood chamber in which she protects them. Most earwigs are tropical or subtropical and many of them have a very restricted distribution. A small number, however, have become widely spread through trade and may be cosmopolitan.

Only a few species have been studied in any detail. Some are mainly phytophagous and are pests of plants in fields and gardens. Others feed to a greater or lesser degree on other insects and their harmfulness as plant pests has to be weighed against their possible value as predators. In most cases, their presence on ships is probably accidental although in some instances they may be associated with populations of stored products pests in the role of predators. Earwigs, therefore, have not assumed much importance in the examination of cargoes or the recording of insect infestations. The Ministry's records show that reported earwig importations number less than 20 a year, of which only about 40 per cent are identified to species.

Anisolabis annulipes (Lucas), see *Euborellia annulipes* (Lucas)

Euborellia annulipes (Lucas)

   (= *Anisolabis annulipes* (Lucas))

Distribution: cosmopolitan.

Interceptions:
   Argentina: bones
   Brazil: Brazil nuts
   Jamaica: coconut shell
   Nigeria: bones, palm kernels, sheanuts
   East Africa: coconuts and coconut shell from Kenya or Tanzania

Mauritius: green ginger, personal effects
India: groundnut cake
Fiji: ginger root
Philippines: nut spillage, desiccated coconut

Biology: this is a world wide species of economic importance and one of the few earwigs that have been studied in detail. It is nocturnal and an omnivorous feeder. As a plant feeder it can damage growing crops and stored vegetables. In the United States, for instance, it is a pest of stored sweet potatoes and will seriously damage the roots of vegetables growing in greenhouses. In India and Israel it is recorded damaging groundnuts by gnawing holes in the pods. It is not, however, always wholly a pest and in its role as a predator it may, in certain circumstances, be classed as a beneficial insect. It is recorded in warehouses and mills where it preys upon grain insects and it is also known as a useful predator of such serious pests as the sugarcane leafhopper, *Perkinsiella saccharicida,* and the banana root borer, *Cosmopolites sordidus.*

The eggs are laid in brood chambers excavated in the soil, sometimes at a depth of 7.5 cm, and within the brood chamber the female cares for the eggs and young nymphs. The lower temperature threshold for nymphal development is just over 16°C and development ranges from 30 days to as much as 280 days according to temperature. In laboratory studies, insects that were reared on meat were found to develop faster, live longer and produce more eggs than those reared on vegetable food. The species is usually wingless but winged forms have been reported in laboratory cultures and in the field. (Klostermeyer, 1942; Neiswandes, 1944; Bharadwaj, 1966; Melamed-Madjar, 1971; El-Husseini and Tawfik, 1972)

Forficula auricularia (L.)

Distribution: this species originated in the Palaearctic region but has now spread to all faunal regions of the world (Brindle, 1970).

Interceptions:
Canada: personal effects
Argentina: bones and bone meal
France: wheat
Denmark: wheat
Greece: currants
Italy: shelled almonds
Korea: gallnuts
Australia: flour

Biology: this is the common earwig, nocturnal and an omnivorous feeder, eating other small insects as well as plant tissue. According to Skuhravy (1960), about 20 to 25 per cent of its food intake is of animal matter, chiefly aphids, and the rest of vegetable matter. Any benefits that derive from its predatory habit are, therefore, considerably outweighed by its harmfulness as a pest of growing plants in fields and gardens. The eggs are laid in brood chambers excavated by the females beneath the surface of the soil and the females care for the eggs and young nymphs within these chambers. In Britain, there is normally one generation a year. (Hickin, 1974; Behura, 1956)

Marava arachidis (Yersin)

Distribution: cosmopolitan. There are two forms, winged and wingless. The winged form is more typically Oriental and Australasian, while Neotropical specimens are all wingless. Wingless specimens are also the most common form in Africa and elsewhere. (Brindle, 1970).

Interceptions:

Argentina: bran and pollards
Brazil: Brazil nuts
Costa Rica: cocoa beans
Guyana: fishmaws
Benin: groundnut cake
East Africa: cottonseed cake, coconuts, capsicums
Ghana: fresh yams, kola nuts
Mozambique: cashew kernels
Nigeria: bones, palm matting, cocoa beans
West Africa: groundnuts, cocoa beans, dunnage
Italy: crushed bones
Bangladesh: beans
Burma: oilcake
India: linseed cake, bones
Malaya: copra
Pakistan: bone meal
Sri Lanka: copra, coconut shell, desiccated coconut, betel nuts
Thailand: rice
Australia: rice
Borneo: illipenuts
Polynesia: copra
Tonga: desiccated coconut

The specimen from Australian rice belongs to the darker, fully winged form which is typical of the Australasian region (A. Brindle, indentifier's comment).

Dictyoptera

This order comprises cockroaches (Blattodea) and mantids (Mantodea). Superficially the members of these two suborders appear very different. Cockroaches are relatively unspecialised in their structure and habits, whereas mantids are highly adapted to a predatory way of life, which has given them a characteristic, somewhat bizarre appearance.

Blattodea

The following brief general account of this suborder is based on information presented by Roth and Willis (1960).

The Blattodea contains many species that are of little or no economic importance. Cockroaches are known from a very wide variety of different habitats, from jungles and forests, beneath bark, in dead wood, in plantations and amongst vegetable debris, in the nests of birds and rodents etc. There are cave dwelling species, desert species and even some aquatic species. The group is best known, however, for the relatively few species which are commensal with man. These are truly omnivorous insects. They will eat not only man's food but also his garbage and excreta, glues and pastes, paper, the size on fabrics and book covers, dried animal and vegetable material, living plant tissue etc. They will eat such a wide variety of substances that it is virtually impossible, despite good housekeeping, to rid a building of all food suitable for cockroaches. Infestations are to be found in homes, restaurants, bakeries, hospitals, laundries, shops, factories and mines as well as on rubbish dumps, in latrines, privies and sewers. They are capable of spreading disease organisms and parasites and cause much unpleasantness by general fouling and contamination of foodstuffs.

Many of the domiciliary species are regular inhabitants of ships and have spread through trade to most of the inhabited regions of the world. In recent years they have also become frequent travellers in aircraft. They are warmth-loving insects and apparently originated in the tropics and subtropics. In the warmer parts of the world, they live and breed out of doors but in the temperate and cooler regions they mostly rely for survival on the sheltered environment provided by man.

Blatta orientalis L., – – common cockroach or oriental cockroach

Distribution: cosmopolitan; of tropical, possibly African or southern Asian, origin but has spread through trade to become one of the most abundant domiciliary cockroaches throughout the world. Although in temperate regions it thrives best in heated buildings, it is not highly susceptible to cold and has been known to occur out of doors in Britain (Solomon and Adamson, 1955; Roth and Willis, 1960). Beatson and Dripps (1972) and Brett (1973) record infestations which had, apparently, survived in the open through several years and they conclude that this species is considerably more cold resistant than has hitherto been realised.

Interceptions:
    United States: dried fruit
    Ghana: cocoa beans, cocoa cake, copra
    Kenya: coconuts
    Nigeria: groundnut cake, bones
    Rhodesia: maize meal
    Sierra Leone: palm kernels
    Tanzania: dried seaweed
    Eire: spaghetti
    Lebanon: lentils
    Turkey: hazelnut kernels
    Burma: rice, groundnut cake
    China: sugared gourds
    Hong Kong: foodstuffs, personal effects
    India: rice, rice bran, groundnut cake, bone meal, gum
    Malaysia: palm kernels, copra, sago flour, illipenuts
    Pakistan: rapeseed meal, guar meal
    Thailand: miscellaneous cargo
    Australia: barley
    Pacific Islands: copra

In spite of its abundance as a domestic pest throughout the world, this species occurs far less often on ships' cargoes than either *Blattella germanica* or *Periplaneta americana*. During the 21 years, 1957 to 1977, it was recorded on 38 occasions.

Biology: this species is one of the commonest cockroaches in bakeries, restaurants, hotels, houses etc. in Britain. The eggs, usually between 12 and 18 at at time, are enclosed in a capsule, or ootheca, and deposited in a warm sheltered situation. The time taken for the eggs to hatch varies from about 40 days at 30°C to 81 days at 21°C and nymphal development ranges from about 300 days at 30°C to as much as 530 days at 24°C. (Ricci, 1964; Guthrie and Tindall, 1968; Hickin, 1974).

The observations of Beatson and Dripps (1972) and Brett (1973) show that development undoubtedly takes place in outdoor conditions in Britain. There is no precise information on the biology of this species at low temperatures but Ragge (1965) suggests that, in unheated conditons in Britain, development may cease altogether during winter and that the insects may take 18 months or two years to become adult. Generally the length of the adult life is around 100 days, a maximum of 273 is recorded.

Blattella germanica (L.) – – German cockroach

Distribution: cosmopolitan. Probably of African origin, this species has spread to become one of the most abundant domiciliary cockroaches throughout the world. In the cooler parts of its range, it relies on the heating of buildings for its survival and within this protected environment it is established as far north as Alaska.

Interceptions: (see Table 4)

Table 4

Records of Blattella germinica L.

Total number of records = 182

Commodity group	Number of records	Geographic region	Number of records
Grain	13	Nearctic	5
Grain products	15	Neotropical	7
Oilseeds	13	Ethiopian	73
Oilcake	30	Northern Europe	1
Cocoa beans	5	Mediterranean	3
Dried fruit	4	Oriental	86
Spices	12	Australasian	7
Coconuts and shell	13		
Animal products	34		
Others	43		

During this period, *Blattella germanica* was the second commonest cockroach species on imported cargoes and was recorded only slightly less often than *Periplaneta americana*.

Biology: this species is one of the commonest pests in homes, restaurants, bakeries etc. in Britain. The eggs, usually between 20 and 45 at a time, are enclosed in a capsule, or ootheca, which is carried around by the female, partly extruded from her abdomen, until the nymphs are ready to hatch. The length of nymphal development varies from about 74 days at 30°C to 170 days at 21°C. Two to three generations a year are possible. Adults usually live for about nine or 10 months, a maximum of 384 days is recorded.

Compared with other domiciliary species, such as *Blatta orientalis* and *Periplaneta* spp., *B. germanica* is a fast breeding cockroach. The life cycle is relatively short and the number of eggs contained within each ootheca relatively large. By being protected within the body of the female, the eggs are less vulnerable to predation and desiccation. These factors contribute towards the high rate of increase and great abundance of this species.

Nauphoeta cinerea (Olivier)

Distribution: Originally an East African species, it has now spread through trade to many tropical countries. Pope (1953) gives an account of it in Brisbane, Australia, and Gurney (1953) records its introduction into commercial buildings in Florida.

Interceptions:
Nigeria: hooves and horns
Burma: groundnut cake, cottonseed cake, rice bran
China: second-hand sacks
Hong Kong: rice sacks, unspecified foodstuffs
India: groundnut meal
Pakistan: rapeseed meal, sesame cake
Sarawak: sago flour
Singapore: copra
Sri Lanka: desiccated coconut

Biology: according to Pope (1953), this is a semi-domestic species in Australia. Adults are found in dwellings but there is no record of them breeding there. They are usually associated with grain stores and fowl feeding pens. In Australia, the average life cycle lasts about a year. The adults are very long lived, a maximum life span of 1185 days is recorded.

Table 5

Records of Periplaneta americana (*L.*)

Total number of records = 213

Commodity group	Number of records	Geographic region	Number of records
Grain	10	Nearctic	6
Grain products	20	Neotropical	17
Oilseeds	19	Ethiopian	83
Oilcake	37	Northern Europe	3
Dried fruit	7	Mediterranean	10
Nuts	7	Oriental	91
Coconuts and shell	8	Australasian	3
Sago flour	5		
Animal products	53		
Others	47		

Periplaneta americana (L.) – – American cockroach

Distribution: cosmopolitan. In spite of its name, it is unlikely to be a native of the New World. It probably originated in Africa, introduced into America with the slave trade (Ragge, 1965).

Interceptions: (see Table 5)

During this period, *Periplaneta americana* was the commonest cockroach species on imported cargoes, although *Blattella germanica* was recorded on almost as many occasions.

Biology: in many parts of the world, this is a domiciliary cockroach although in Britain it is not normally encountered in dwellings, being found more often in commercial buildings around ports. It is particularly common on ships.

The eggs, usually 14 to 16 at a time, are enclosed in capsules or oothecae which are carried around by the female before being deposited. The incubation period varies from 30 days at 30°C to 88 days at 17 to 18°C. The length of nymphal development is very variable, even amongst individuals reared under identical conditions. It may be as short as four to five months at 25 to 30°C while, in unheated conditions, it may take as long as two to three years. Usually the life cycle takes about a year. The adults are long-lived, a maximum life span of 1502 days is recorded. (Pope, 1953; Ragge, 1965; Guthrie and Tindall, 1968)

In addition to an omnivorous scavenging way of life, typical of cockroaches, this species is recorded as a predator of other insects. (Bowden and Phipps, 1968; Cooke, 1968)

Periplaneta australasiae (Fabricius) – – Australian cockroach

Distribution: cosmopolitan. In spite of its name, this species probably originated in Africa or southern Asia. Pope (1953) suggests that Fabricius may have used the name *australasiae* to mean 'of southern Asia' and not to refer specifically to Australia.

Interceptions:
Brazil: Brazil nuts
Guyana: fishmaws
Nigeria: groundnuts
China: cassia bark
India: groundnut cake, cottonseed cake
Malaya: tapioca
Sarawak: sago flour
Thailand: fabric
Australia: panicum and millet, rice, nickel

This species is infrequently recorded on imports. During the 21 years, 1957 to 1977, it was reported on only 12 occasions.

Biology: the eggs, up to 28 at a time, are enclosed in capsules or oothecae which are carried around by the female for about 24 hours before being deposited. The eggs usually take seven to eight weeks to hatch in summer, but up to 23 weeks at less favourable times of year. Nymphal development usually takes about a year. The average life span of adults is 18 months, a maximum of 937 days is recorded. (Pope, 1953, Ragge, 1965)

Psocoptera – – psocids, booklice

These are small primitive insects, either winged or wingless, that feed mainly on moulds, lichens and other vegetable material in damp situations. Some have been recorded eating the eggs of other insects. Several species are to be found indoors in domestic and storage premises, sometimes becoming pests, contaminating and damaging foodstuffs in larders and warehouses. They are particularly associated with old damp buildings but may also become numerous in newly built houses during the time that the plaster is drying out. (Hickin, 1974)

Lachesilla pedicularia (L.) – – cosmopolitan grain psocid

Distribution: cosmopolitan.

Interception:
    Turkey: sultanas

Lepinotus patruelis Pearman – – black domestic psocid

Distribution: Europe, introduced into North America, California, in granaries etc. (Finlayson, 1949).

Interceptions:
    Eire: milk powder
    France: walnuts

Liposcelis bostrychophilus Badonnel – – stored product psocid

Distribution: cosmopolitan.

Interceptions:
    Argentina: maize, wheat
    Nigeria: ginger
    Somali: bones
    India: coriander seed

Biology: this is a well known domestic and storage pest. In particular it is a secondary pest of maize, occurring when there has been previous damage by other insects. It is also a pest of live insect cultures in laboratories and has been recorded eating the eggs of various lepidopterous and coleopterous grain pests. (Shires, 1976)

Liposcelis entomophilus (Enderlein)

Distribution: cosmopolitan, in stored products.

Interceptions:
    Canada: wheat
    Argentina: wheat
    Iraq: barley
    Thailand: rice

Liposcelis mendax Pearman

Interception:
    Australia: millet

Liposcelis paetus Pearman

 Distribution: cosmopolitan, in stored products.

 Interceptions:
  Argentina: wheat, oats
  South Africa: kaffir corn
  Sudan: cottonseed
  Tanzania: maize meal
  Iraq: barley
  Burma: rice
  India: groundnut cake, linseed cake, crushed bones
  Pakistan: crushed bones

Psoquilla marginepunctata Hagen

 Distribution: originally from America but fairly frequently imported into Europe (Badonnel, 1943). Forsyth (1966) records it in stored cocoa in Ghana.

 Interception:
  Brazil: Brazil nuts

Troctes spp., see *Liposcelis* spp.

Hemiptera

The Hemiptera are sucking insects whose mouthparts, in the immature nymphal stages as well as in the adults, are modified to form a piercing rostrum or beak. They feed on liquids only. Several species have been recorded on stored products, on which they occur either as predators on populations of insect storage pests or as accidental strays.

The order is divided into two suborders, the Homoptera and the Heteroptera. The salient difference between these two groups lies in the structure of the forewing. In the Homoptera, the forewing is of the same texture throughout and the wings, when at rest, are held angled, like a roof, over the body. The Heteroptera are characterised by having the forewing divided into a basal, thickened area and an apical membranous area. The wings are held flat over the body, with the membranous areas overlapping. Wingless forms occur in both the Homoptera and Heteroptera.

Hemiptera — Homoptera

The Homoptera are plant feeders, piercing the plant tissues and sucking the juices. Among them are the cicadas, leafhoppers, froghoppers, aphids and scale insects and they include some of the most troublesome plant pests. In spite of being extremely abundant and widespread throughout the world, members of the Homoptera are rarely identified or recorded from imported cargoes by the Ministry's inspectorate. Aphids are occasionally found on food cargoes but tend to be fragile and easily damaged during collection and I have no records of any being identified to species.

Hemiptera — Heteroptera

The Heteroptera, which are the true bugs, show a remarkable diversity of form and habit. Many are plant feeders and may be serious plant pests, others are voracious predators on other insects, even, as in the case of the Belostomatidae, on small vertebrates, while some are adapted to sucking the blood of warm-blooded animals.

The great majority of Hemiptera recorded from imports belong to the Heteroptera.

Anthocoridae

Commonly known as flower bugs, the members of this family are predacious. Although a few common species are often found on flowers, most live in more sheltered places such as under bark, in decaying vegetable matter, in beetle galleries in bracket fungi, stored products, nests of birds and rodents and in several kinds of epiphytes (Herring, 1967). A few species in the genera *Lyctocoris* and *Xylocoris* occur fairly regularly on imported food cargoes where they prey on the associated insect fauna.

Lyctocoris campestris (Fabricius) – – stack bug

> *Distribution:* a cosmopolitan species that has been widely distributed by trade throughout the world. Cornes (1973) records it in Nigeria and McFarlane (1963) records it in Jamaica. According to Woodroffe (1953) it is the commonest predator found in birds' nests and in warehouses in Britain.

Interceptions: (Table 6)

Table 6

Records of Lyctocoris campestris *(Fabricius)*

Total number of records = 61

Commodity group	Number of records	Geographic region	Number of records
Grains and grain products	12	Nearctic	2
Oilseeds and oilcake	17	Neotropical	9
Nuts	6	Ethiopian	16
Dried fruit	6	Northern Europe	0
Animal products	7	Mediterranean	15
Others	13	Oriental	19
		Australasian	0

This is the Anthocorid most commonly recorded from imported cargoes.

The single specimen from the Nearctic region was found on flour from the southern United States. From the Neotropical region specimens came from Argentina, Brazil and Jamaica.

Although the species occurs in northern Europe and the Ministry holds numerous records from British mills, granaries and stores, we have no certain record of its importation from the northern countries of Europe. Five specimens were imported from France, three on cargoes of walnuts and two on cargoes of maize. In the absence of information on the port of loading or origin of these cargoes, I have counted these records as from the Mediterranean region since they seem most likely to have come from the southern part of France. Other Mediterranean records were from cargoes from Algeria, Greece, Iran, Lebanon and Portugal.

From the Ethiopian region, specimens came from Nigeria, Rhodesia, Kenya, Tanzania and Sudan, and from the Oriental region, from India, Bangladesh, Pakistan, Burma, Malaya and Thailand.

Biology: this species is found in a variety of habitats, including vegetable debris, hay stacks, granaries, warehouses, stables, fowl houses etc. Woodroffe (1953) records it as a predator of the larvae of house moths. It will also suck the blood of warm-blooded animals and has been recorded infesting the clothing and bedding of man (Southwood and Leston, 1959). Stys (1973) records a case in Czechoslovakia of an infestation being introduced into flats with hay mattresses.

Piezostethus spp., see *Xylocoris* spp.

Xylocoris afer (Reuter) – – African store bug

Distribution: tropical and subtropical Africa, Madagascar, Reunion, New Guinea, large part of tropical America, Antilles. Its presence in the Palaearctic region has not yet been proved (Carayon, 1972).

Interceptions:

Argentina: wheat
Brazil: Brazil nut residues
Jamaica: coconut shell
Nigeria: cocoa beans, soya beans, groundnuts, ginger root
Kenya or Tanzania: coconuts and coconut shell
Turkey: wheat
Burma: rice bran
India: rice, rice bran
Indonesia: dried ginger
Malaya: sandalwood

Xylocoris flavipes (Reuter) – – cosmopolitan cereal bug

Distribution: a widely distributed species, native of Africa and the Far East it does not occur naturally in the United Kingdom (P.S. Broomfield, identifier's comment). Known from all warmer parts of the world (Jay, Davis and Brown, 1968).

Interceptions: (Table 7)

The single specimen from the Neotropical region was from Argentinian hoof and horn meal. One specimen came from Rumanian maize, from the most southerly edge of the northern European region.

Biology: This species is particularly associated with stored products. It preys upon a wide variety of storage pests although those species (e.g. *Sitophilus* spp.) that spend their entire larval life within the grains, escape attack. Studies by Jay, Davis and Brown (1968), Le Cato and Davis (1973), Press, Flaherty and Arbogast (1974) and Arbogast (1976) suggest that this species is

Table 7

Records of Xylocoris flavipes *(Reuter)*

Total number of records = 43

Commodity group	Number of records	Geographic region	Number of records
Grain and grain products	14	Nearctic	0
Oilseeds and oilcake	19	Neotropical	1
Others	10	Ethiopian	13
		Northern Europe	1
		Mediterranean	3
		Oriental	24
		Australasian	1

a useful predator and may successfully prevent the build up of large populations of insect pests in stored products. It is a general predator, however, and when it occurs in association with more specialised parasites, such as *Bracon hebetor*, it will prey as readily on the *Bracon* as on the pest species, thus tending to reduce rather than enhance the effectiveness of their combined biological control.

Awadallah and Tawfik (1972) studied its life cycle at room temperature in Egypt and found that the average developmental period of the egg and nymphal stages are four and 13 days respectively.

Xylocoris galactinus (Fieber) – – hot-bed bug

Distribution: Europe, Africa, Asia, North and South America (Herring, 1967). Carayon (1972) states that it is widespread in the Palaearctic region and often found in cargoes but that it is not proved to be cosmopolitan. It occurs in Britain, where it thrives in manure heaps, hot beds, stable straw, grain bins etc. where the temperature is quite high.

Interceptions:
  Greece: carobs
  Ghana: cocoa beans
  Mozambique: cashew nut kernels
  Nigeria: cocoa beans, bones
  India: gum karaya
  Malaya: coconut shell
  Thailand: tapioca

Biology: this has been studied by Tawfik and El-Husseini (1971) in Egypt. In the laboratory, at about 26°C and 64 per cent relative humidity and with mites and young housefly larvae as food, the complete life cycle took about 22 days. Southwood and Leston (1959) found that the species thrived at 27°C but could tolerate up to 42°C, although above 32°C, increase in humidity is detrimental.

Miridae

These bugs are called capsid bugs after the former name of the family, Capsidae. It is a very large family and its members are mainly plant feeders, often doing serious damage to crops. Some, however are at least partially predatory and a few entirely so (Southwood and Leston, 1959). The only mirids recorded here are of the genus *Fulvius*, which belongs to the subfamily Cylapinae comprising predatory or partially predatory species (Woodroffe and Halstead, 1959).

Fulvius brevicornis Reuter

   Distribution: Christmas Is., Venzuela, Martinique, Sri Lanka, Taiwan, Africa, Cuba, St. Thome, Burma, East Indies, Asia, Brazil, North America, Bonin Is., Mariana Is. (Carvalho, 1956).

   Interceptions:
      Brazil: Brazil nuts, Brazil nut residues, logs near Brazil nuts
      East Africa: logs
      Nigeria: copra, logs
      West Africa: mouldy logs
      Malaya: sandalwood
      Sri Lanka: coconut shell

   There have been more importations from Brazil than from any other country. Woodroffe and Halstead (1959) record its survival and breeding in a stack of Brazil nuts under storage conditions in Britain, at temperatures varying around 20°C

Nabidae

These are damsel bugs, which are active predators often playing a part in the biological control of insect pests. The eggs are laid embedded in grass stems or in other plant tissue (Southwood and Leston, 1959).

Pagasa fusca (Stein)

   Distribution: North and South America (R. J. Izzard, identifier's comment).

   Interception:
      Argentina: pollards

Reduviidae

This family includes the assassin bugs which are large and active hunters, catching and sucking live prey. They are characterised by the short curved rostrum which they insert into the intersegmental membrance of the prey's body as it is clasped in their jack-knife fore limbs (Ryckman and Ryckman, 1967). The family also contains some blood-sucking bugs in the subfamily Triatominae which feed on the blood of mammals, birds and reptiles. The Triatominae do not feature amongst the species listed here.

Amphibolus venator (Klug)

   Distribution: widespread in the tropics (various authors).

   Interceptions: (Table 8)

Table 8

Records of Amphibolus venator *(Klug)*

Total number of records = 207

Commodity group	Number of records	Country	Number of records
Grain and grain products	12	Ethiopian region	
Oilseeds	5	(Aden, Ethiopia,	
Oilcake	158	Gambia, Nigeria,	31
Animal products	23	Portuguese Guinea,	
Others	9	Senegal, Sudan,	
		Tanzania, Zaire	
		Burma	124
		India	35
		Malaya	3
		Pakistan and Bangladesh	14

In spite of the fact that this is the most frequently recorded of all Hemiptera species on imports, it was recorded from relatively few countries. It is interesting to note, however, that the pattern of importations of *Amphibolus venator* reflects very closely that of the khapra beetle (*Trogoderma granarium*) which, according to Tembe and Surlikar (1968) is its preferred prey. The majority of interceptions came from oriental cargoes, with Burmese oilcake and, to a lesser extent, Indian oilcake accounting for almost all of them (Aitken, 1975).

Biology: this predatory species is mainly associated with stored products and feeds voraciously on all stages of a wide variety of coleopterous and lepidopterous stored products pests. It is particularly abundant with the khapra beetle (*Trogoderma granarium*) and this species appears to be its preferred prey. It is regarded as a generally beneficial insect in stores and its usefulness is not confined to the control of *Trogoderma*. It has, for instance, been observed to reduce infestations of *Ephestia cautella* and *Alphitobius diaperinus* and to feed abundantly on the adults of *Rhyzopertha dominica*. It will breed readily at temperatures from 21° to 32°C and from 40 to 70 per cent relative humidity but is killed by temperatures below 13°C or above 40.5°C. The length of life cycle varies widely even under identical conditions and nymphal developments ranging from 41 to 165 days are recorded; Pingale (1954), Tembe and Surlikar (1968), Surlikar and Tembe (1969), Hussain and Aslam (1970).

Peregrinator biannulipes (Montrouzier and Signoret)

Distribution: widespread in the tropics. Wygodzinsky and Usinger (1960) record it in Mexico, Central America, Cuba, Reunion Is., Rodrigues, Philippines, New Guinea, Amboina, New Caledonia, Fiji and Micronesia. Le Pelley (1959) records it in Kenya and Uganda, Forsyth (1966) in Ghana, Cornes (1973) in Nigeria, Hinckley (1963) in Fiji. The Ministry also holds records of the species collected in Pakistan and Senegal by Dr. J. A. Freeman.

Interceptions:

Benin: groundnut cake
Nigeria: ginger, matting near oilcake
Kenya or Tanzania: coconuts and coconut shell
Zanzibar: cassava root
East Africa: cassava root, cottonseed cake
Burma: cottonseed cake
Borneo: illipenuts
India: rice bran, groundnut cake, bones
Indonesia: dried ginger
Sarawak: illipenuts
Singapore: illipenuts
Sri Lanka: desiccated coconut

Reduvius personatus (L.) – – masked hunter bug

Distribution: A European species introduced into the United States, where its range extends from the eastern states to Arizona and California (Ryckman and Ryckman, 1967). In Britain, it does not occur north of Lancashire, nor in Wales, Scotland or Ireland (Southwood and Leston, 1959).

Interception:

Algeria: wheat bran

Biology: this is a general predator in warehouses, dwellings, outhouses etc., predatory on silverfish, flies, bed bugs, harvestmen etc. It is almost always found in buildings and in association with man but there are a few recorded instances of its occurrence in the nests of bats, rodents and birds. It is often found in association with bed bugs (*Cimex lectularius*) and the nymphs camouflage themselves with dust and debris, earning the common name of masked bug or masked hunter bug. Its life cycle has not been studied in detail but observations in America show that some individuals may take two years to develop, spending two winters as nymphs. (Woodroffe, 1953; Southwood and Leston, 1959; Ryckman and Ryckman, 1967).

Lepidoptera

The Lepidoptera comprise the butterflies and moths and are probably the best known and most ardently collected and observed of all the insects. Although the terms 'butterfly' and 'moth' are in such common usage, there is little difference between them. The two most commonly known characteristics of butterflies are their clubbed antennae and the absence of any frenulum as a coupling mechanism for the wings. These characters, however, are also to be found amongst some of the moths. Of greater value in identifying the butterflies are characters of wing venation and the development of the palpi.

The larvae or caterpillars of the majority of the Lepidoptera feed on living plants. Many are harmless feeders on wild plants but some are voracious pests of cultivated plants and food crops, capable of doing a great deal of damage in fields and gardens. A few are wood borers, some in rotting wood but others in the living wood of trees and bushes. Amongst the crop pests are some moths (e.g., *Sitotroga cerealella* on grain and *Phthorimaea operculella* on potatoes) which are able to continue their attack on the stored crop after harvest and are, therefore, both field and storage pests. Others, notably amongst the Pyralidae, are exclusively storage pests, living on dried stored produce in warehouses, granaries and mills. The larvae of some moths are able to feed on materials of animal origin, either as part of an omnivorous diet, as in the house moths (Oecophoridae) or as a wholly animal diet as in the clothes moths (in the Tineidae).

In the Lepidoptera it is almost always the larvae that cause damage, either directly by their feeding or indirectly by contaminating their food or habitat with webbing or excreta. The adults either do not feed at all or take only liquid food. Within the Noctuidae, however, are some species in which the feeding of the adults is of some significance. These are the fruit-piercing moths which as adults are capable of doing damage by piercing the skins of various fruits in order to suck the juices while their larval stages feed on plant foliage; also some eye-sucking moths suck the lachrymatory fluids from the eyes of animals and man, causing irritation.

Of the species of Lepidoptera recorded here, only one is a butterfly and that is a field pest belonging to the Pieridae. All the rest are moths, wide ranging in their economic importance, from harmless strays and unimportant scavengers to some of the most serious of insect pests of growing crops and stored food.

Gelechiidae

Pectinophora gossypiella (Saunders) – – pink bollworm

(= *Platyedra gossypiella* (Saunders))

Distribution: in all the major cotton producing countries throughout the world, with the exception of the Soviet Union, South Africa, Nicaragua and Peru (Lukefahr, Noble and Martin, 1964). It is noteworthy, however, that the Ministry recorded it from a cargo of cottonseed meal imported from Peru in 1960 and this may indicate that the species does, in fact, occur there.

Interceptions:

Peru:	cottonseed meal
Morocco:	cottonseed
Spain:	cottonseed
Angola:	coffee beans
Ivory Coast:	cottonseed
Nigeria:	cottonseed, cocoa beans, coffee beans
Sudan:	cottonseed
Togoland:	cottonseed
West Africa:	cottonseed

Biology: this is the pink bollworm and is a major pest of cotton, infesting the flowers, buds and bolls of the growing plant and feeding especially on the developing seeds. Although cotton is its preferred host, it will also feed on many other plant species, with okra (*Hibiscus esculenta*) being preferred next to cotton (Chapman, Noble, Robertson and Fife, 1960).

Taher, El-Sayed and Abdel-Rahman (1960) found that the life cycle ranged from 75 days at 18°C and 42 days at 25°C to 25 days at 35.5°C. No eggs were produced at 35.5°C and only a few at 18°C. Some larvae enter diapause in which they survive the winter, the onset of diapause being stimulated by a decrease in the day length. No diapause occurs in tropical areas where day length remains constant throughout the year but the incidence of diapause progressively increases at latitudes remote from the equator (Lukefahr, Noble and Martin, 1964).

Platyedra gossypiella (Saunders), see *Pectinophora gossypiella* (Saunders)

Sitotroga cerealella (Olivier) – – Angoumois grain moth

   Distribution: throughout most warm and warm-temperate regions of the world. It is not normally considered to overwinter under unheated conditions in Britain, although Solomon and Adamson (1955) observed a few larvae to survive fairly rigorous winter conditions in this country. Observations in Austria suggest that the species may be able to survive the winter there in unheated buildings and even to establish infestations in field grain, possibly as a result of a succession of mild winters (Faber, 1978).

   Interceptions: (see Table 9)

Table 9

Percentage of cargoes found infested with Sitotroga cerealella (Olivier) *where 10 or more cargoes were examined.*

	Nearctic	Neotropical	Ethiopian	Northern Europe	Mediterranean	Oriental	Australasian	Total nos. of cargoes
Maize	<1	32	2	1	0	0	0	284
Wheat	<1	12	0	<1	<1	0	<1	140
Rice	<1	<1	<1	0	<1	<1	<1	12
Other grains	<1	4	<1	0	<1	<1	<1	40
Grain products	0	3	<1	0	<1	0	<1	47
Others	<1	<1	<1	<1	<1	<1	<1	102
Total nos. of cargoes	56	408	36	41	15	19	50	625

   The majority of records were from South American cargoes, particularly from Argentina, which accounted for 374 infestations (or 60 per cent of the total). Most of these occurred during the first eight years of the period (1957 to 1964) during which 52 per cent of Argentinian maize and 17 per cent of Argentinian wheat cargoes were found to carry the species. During the following 13 years (1965 to 1977) these proportions had fallen to five per cent and less than one per cent respectively. Of the 38 records from northern Europe, 19 were from French maize and 15 from Rumanian maize, and almost all of these occurred during the first nine years of the period.

   Biology: this is the Angoumois grain moth, a serious pest of all kinds of stored grain in many parts of the world. It will also attack ripening grain in the fields and has been reported feeding on grasses, bran, dried fruits, chick peas, cowpeas, bamboo seeds etc. (Shahjahan, 1974). The eggs are laid on the outside of the grains and the newly hatched larvae bore their way into the grains within which they spend their entire larval and pupal development.

Optimum conditions for the development of all stages is around 30°C and from 65 to 80 per cent relative humidity at which the life cycle from egg to adult may take as little as 25 to 29 days. At temperatures above 30°C the life cycle takes longer (29 to 36 days at 35°C) and development is impossible at 40°C. The longest life cycle (113 days) was recorded at 17°C. Development tends to take longer on maize than on wheat or rice. (Vukasovic, 1940; Hammad, Shenouda and El-Deeb, 1969; Grewal and Atwal, 1969; Boldt, 1974).

Oecophoridae

Endrosis sarcitrella (L.) – – white-shouldered house moth

(= *Endrosis lactella* Denis and Schiffermüller)

Distribution: widespread in temperate regions, incapable of living in the tropics. Howe and Freeman (1955) consider that specimens found on imports from West Africa probably originated from infestations living in residues in the holds of ships.

Interceptions: (see Table 10)

Table 10

Records of Endrosis sarcitrella (L.)

Total number of records = 164

Commodity group	Number of records	Geographic region	Number of records
Grain	97	Nearctic	65
Grain products	28	Neotropical	13
Oilseeds	2	Ethiopian	13
Oilcake	6	Northern Europe	42
Dried fruit	7	Mediterranean	5
Nuts	4	Oriental	13
Pulse and pulse products	7	Australasian	13
Animal products	4		
Others	9		

The majority of records were from the temperate zones of North America and northern Europe, where the species occurs naturally. There were also, however, infestations on cargoes from some of the warmer regions of the world — from Argentina, Mexico and Chile in the Neotropical region, from Kenya, Tanzania, Mozambique, Nigeria and South Africa in the Ethiopian region, from Greece, Turkey, Spain and Italy in the Mediterranean region and from India, Pakistan, Sri Lanka, Afghanistan and Burma in the Oriental region. The species is known to be associated with moulds and residues in the holds of cargo ships (Hurlock, 1963, 1964) and it is likely that some of these records, particularly those associated with tropical cargoes, represent cross-infestations from such residues.

Biology: this is the white-shouldered house moth. Woodroffe (1951a) found that at 90 per cent relative humidity the life cycle, from egg to adult, ranged from 62 days at 25°C to 235 days at 10°C. at 80 per cent relative humidity complete development was achieved only at 20 and 25°C and development failed at all temperatures when the relative humidity dropped to 70 per cent. No diapause was observed and the larvae pupated within a fortnight of becoming full grown.

Endrosis sarcitrella is often found in association with *Hofmannophila pseudospretella* and shares with it a requirement for a relative humidity of at least 80 per cent. In other respects, however, the life histories of these two species are dissimilar. *Endrosis* develops relatively more quickly than *Hofmannophila* and lacks the diapause that prolongs the life cycle of that species. *Endrosis* is, therefore, able to produce several generations a year under normal warehouse or domestic conditions, compared with the single generation which is usual for *Hofmannophila*. *Endrosis* tends to be less common in houses than *Hofmannophila* and is more often associated with stored food products than with fabrics and household effects. Mixed populations of both species infesting a large granary at a port in Scotland have been studied recently by Cole and Cox (1981).

Hofmannophila pseudospretella (Stainton) – – brown house moth

Distribution: widespread in temperate regions, incapable of breeding in the tropics. Howe and Freeman (1955) consider that specimens found on imports from West Africa probably originated from infestations living in residues in the holds of ships.

Interceptions: (see Table 11)

Table 11

Records of Hofmannophila pseudospretella *(Stainton)*

Total number of records = 125

Commodity group	Number of records	Geographic region	Number of records
Grain	64	Nearctic	41
Grain products	31	Neotropical	13
Oilseeds	1	Ethiopian	11
Oilcake	6	Northern Europe	37
Dried fruit	2	Mediterranean	7
Nuts	6	Oriental	6
Pulse and pulse products	6	Australasian	10
Animal products	7		
Others	2		

As with *Endrosis* the majority of records were from the temperate zones of North America and northern Europe, where the species occurs naturally. There were also infestations on cargoes from some of the warmer regions of the world — from Argentina and Colombia in the Neotropical region, from Kenya, Tanzania, Nigeria, Ghana and South Africa in the Ethiopian region, from Spain, Portugal, Italy, Greece and Lebanon in the Mediterranean region and from India, Burma and Indonesia in the Oriental region. The species is listed by Hurlock (1963, 1964) as an inhabitant of mouldy residues in the holds of cargo ships and it is likely that some of our records, particularly those associated with tropical cargoes, represent cross-infestations from such residues.

Biology: this is the brown house moth the larvae of which can feed on a very wide range of substances of animal or vegetable origin. Höller (1965) demonstrated their ability to digest wool, particularly in their later stages. *Hofmannophila* has a particular importance as a pest of fabrics, household furnishings, carpets, underfelts, residues and fluff. It is probably now established in most private houses in Britain (Hickin, 1974). Woodroffe (1953) lists it as probably the commonest insect species in birds' nests in Britain and nests built in roof spaces and eaves of houses are probably one of the main sources of domestic infestations.

Woodroffe (1951 b) found that larval development was possible only when the relative humidity was at least 80 per cent. Temperatures over 25°C had a progressively adverse effect on larval growth. The duration of the life cycle is very variable. At 90 per cent relative humidity the larval feeding period ranges from about 71 days at 25°C to 145 days at 13°C and more than 180 days at 10°C. After feeding, the larvae enter diapause which may prolong the over-all development to as much as 266 days at 25°C and 440 days at 20°C.

Hofmannophila is frequently found in association with *Endrosis sarcitrella* and shares with it a requirement for a relative humidity of at least 80 per cent. In other respects, however, the life histories of these two species are not similar. Not only do the larvae of *Hofmannophila* develop more slowly, but the addition of a diapause at the end of the feeding period further prolongs its life cycle so that only a single annual generation is produced, compared with several generations of *Endrosis*.

Pyralidae

This is a vast family, comprising over 20 thousand species divided into 8 sub-families. Five of these sub-families, the Galleriinae, Crambinae, Phycitinae, Pyralinae and Pyraustinae, contain species of economic importance some of which are amongst the most destructive pests.

Galleriinae

Aphomia gularis (Zeller), see *Paralipsa gularis* (Zeller)

Corcyra cephalonica (Stainton) – – rice moth

Distribution: throughout the hot regions of the world. Freeman (1962), in a survey of flour mills, found that *C. cephalonica* occurred in mills in hot humid regions where temperatures are over 22°C and the relative humidity exceeds 70 per cent and that it replaces *Ephestia kuehniella* as the dominant mill moth in these regions.

Interceptions: (see Table 12)

Insect Travellers

Table 12

Percentage of cargoes found infested with Corcyra cephalonica *(Stainton) where 10 or more cargoes were examined.*

	Nearctic	Neotropical	Ethiopian	Northern Europe	Mediterranean	Oriental	Australasian	Total nos. of cargoes
Wheat	<1	0	0	<1	0	0	0	3
Maize	<1	<1	1	0	0	9	0	23
Rice	<1	2	23	0	<1	16	<1	354
Others grains	<1	<1	4	0	0	0	<1	26
Wheat products	<1	<1	6	<1	1	2	0	42
Maize products	0	0	7	0	0	4	0	126
Rice products	<1	8	20	0	11	22	0	499
Oilseeds and copra	0	1	7	0	<1	3	<1	554
Oilcake	<1	<1	26	0	<1	16	<1	2688
Cocoa beans	0	6	1	0	0	<1	16	166
Dried fruit	0	0	<1	0	1	0	<1	77
Nuts	0	<1	0	<1	<1	1	0	36
Pulse and pulse products	0	0	2	0	<1	3	<1	88
Animal products	<1	<1	<1	0	0	<1	0	33
Illipenuts	0	0	0	0	0	17	0	25
Others	0	<1	1	0	<1	<1	<1	202
Total nos. of cargoes	18	153	2341	4	114	2195	117	4942

During this period, there was a marked reduction in the incidence of this species on imports, from 512 records in 1957 to 63 records in 1977.

From Africa, oilseeds (particularly groundnuts) and oilcake (particularly groundnut cake) were by far the most common sources of this species, with Kenya, Tanzania and Nigeria being the most frequent exporting countries. Howe and Freeman (1955) studied *C. cephalonica* on West African imports during the period 1942 to 1952 and found that it was particularly associated with groundnut cargoes. They found a decrease in the number of infestations, from 58 per cent of cargoes at the beginning of their period, to 43 to 46 per cent at the end. A comparison with the present survey shows that the downward trend between 1942 and 1952 has been continued, even accelerated during the period up to 1977. In 1957 the percentage of West African groundnut cargoes affected was 42 per cent, only slightly below the level for 1952. From 1958 onwards, however, there was a very marked decrease, so that by 1975 only 8 per cent of cargoes was affected and none at all in 1976 and 1977.

Oilcake from the Oriental region, particularly from Burma was also a frequent carrier of the species. The incidence on Burmese oilcake was rather irregular and varied considerably from year to year. Nevertheless, an over-all reduction in infestation levels has occurred. During the first half of the period, from 1957 to 1966, 60 per cent of these cargoes were affected but in the second half, up to 1974, the percentage was 38. No Burmese oilcake was examined in 1975, 1976 or 1977. It was from the Oriental region that most of the rice and rice products infestations came. On rice, mainly from Burma and Thailand, a very clear reduction occurred — from 60 per cent of Burmese rice and 43 per cent of Thai rice cargoes affected in 1957 to 0 per cent in 1976 and 1977. Indian rice bran cargoes also showed a significant decrease, from 40 per cent in 1957 to nine per cent in 1977. Burmese rice bran, however, maintained a higher incidence of *Corcyra* throughout the period, reducing only from 67 per cent in 1957 to 38 per cent in 1973 (no cargoes were inspected from 1974 onwards).

Biology: this moth is one of the most important pests of flour mills in warmer regions. The larvae produce a considerable quantity of silk which, in heavy infestations, can form dense masses in mill machinery and over the surface of food. On whole grains, it is regarded as a secondary pest, attacking grains that have already been damaged by other insect pests and rarely attacking whole undamaged grains. It is known to feed on quite a wide range of stored products but thrives best on a cereal diet. Broken or coarse milled grains are attacked most readily whereas fine milled flours as well as whole grains are relatively unsuitable as food.

Under laboratory conditions, the average larval development on corn meal ranges from 36 days at 25.7°C to 117 days at 18°C. Larval development on finely sifted wheat flour, however, was considerably slower — 50 days at 25.5°C. In Egypt there are normally about six generations a year on corn meal and five generations a year on wheat flour. Rao (1954), Kamel and Hussanein (1967) and Punj (1967). Limits for development from egg hatch to adult emergence were about 17°C and 35°C at 70 per cent relative humidity, with highest survival and most rapid development occuring at 30–32.5°C. Eggs failed to hatch below 17.5°C (Cox *et al* (1981a)).

Paralipsa gularis (Zeller) – – stored nut moth

(= *Aphomia gularis* (Zeller)

Distribution: this species almost certainly originated in South East Asia where it occurs widely, often as a pest of rice. Elsewhere in the world it has been spread by trade to Europe and North America, where it occurs locally in warehouses and manufacturing premises where dried fruit or nuts are stored. There are no published records as yet from the southern hemisphere. It is essentially a sub-tropical and warm-temperate species, being rare in the tropics and maintaining itself towards the northern limits of its range (e.g. in northern Britain and Sweden) in the store rooms of heated factories. (Smith, 1956; 1965)

Interceptions: (see Table 13)

Records from the northern European region were all from France, mainly from cargoes of shelled walnuts from Bordeaux. French walnuts were by far the most common source of this species and accounted for 45 of the 68 records. An earlier survey of this species on French walnuts during the three years July 1955 to June 1958 was made by Smith (1960) and he found that 43 per cent of cargoes were affected during that period. This present analysis, however, shows that there has been a very sharp decline in the incidence of the species since then. In 1957, 38 per cent of French walnut cargoes were

Table 13

Records of Paralipsa gularis *(Zeller)*

Total number of records = 68

Commodity group	Number of records	Geographic region	Number of records
Grain	9	Nearctic	0
Grain products	1	Neotropical	0
Oilcake	1	Ethiopian	1
Nuts	56	Northern Europe	47
Others	1	Mediterranean	18
		Oriental	2
		Australasian	2

affected. During 1958 this had dropped to 19 per cent and by 1965 to less than one per cent. No *Paralipsa* was recorded at all from this source during the period 1966 to 1977.

From the Mediterranean region the species was most frequent from Italy, mainly on rice, and was also imported from Turkey and Morocco. From the Oriental region it was imported once on Indian walnuts and once on Indonesian pepper. I have been unable to find any published records of this species in southern Africa and it seems likely, therefore, that the single record from Tanzanian cottonseed cake represents a cross-infestation from some other source.

Biology: this species is known as a pest of grain, seeds, stored nuts and dried fruit and it can cause serious losses if it becomes established in heated factories, where valuable commodities such as almonds may be attacked. In southern India it occurs as a pest of tamarind pods ripening on the trees and this may represent its natural habitat.

In the laboratory, at 70 per cent relative humidity and on a diet of groundnuts, larvae develop within the temperature range 15°C to 33°C, with an optimum temperature of about 31°C. When bred at relatively low temperatures or in crowded conditions, larvae enter diapause, in which condition they are able to withstand temperatures as low as –3°C for up to three months (Smith, 1965).

Phycitinae

This sub-family is the largest in the Pyralidae and is of particular importance since it contains some of the most widespread and serious of all the lepidopterous pests of stored products (notably in the genera *Ephestia* and *Plodia*). Many of the Phycitinae, however, are field pests or relatively harmless plant feeders.

Anagasta kuehniella (Zeller), see *Ephestia kuehniella* Zeller

Cadra spp., see *Ephestia* spp.

Ectomyelois ceratoniae (Zeller) – – locust bean moth

(= *Myelois ceratoniae* Zeller)

(= *Spectrobates ceratoniae* (Zeller))

Distribution: although associated in particular with produce from the Mediterranean region, this species is found in most tropical parts of the world, in South America, the West Indies, Hawaii and it has been spread by trade to parts of southern Asia (Balachowsky, 1972). Catling (1970) describes it as a pest of citrus in South Africa and Michael (1968) reports its occurrence in Western Australia.

Interceptions:

Trinidad:	tamarinds
Algeria:	dates
Crete:	carobs
Cyprus:	carobs, carob meal
Greece:	carobs
Italy:	almonds
Majorca:	almonds
Portugal:	almonds
Spain:	raisins, almonds, carobs
Tunisia:	dates

Biology: this species is known particularly as a pest of carobs, citrus fruits, dried fruit and almonds in the Mediterranean region. It has, however, been recorded on quite a wide range of other fruits including pomegranates, apples, figs, dates, olives and the pods of tamarinds and *Acacia*. Infestation starts in the field but may continue in the dried and stored crop after harvest.

On citrus, it seems that the presence of mealybugs and the honeydew secreted by them may play a significant part in attracting the moth to the fruit and encouraging oviposition. In this respect it resembles another Phycitinine moth *Cryptoblabes gnidiella* — in fact at one time the damage caused by *Ectomyelois* was wrongly attributed to the less destructive *Cryptoblabes*.

Except on some citrus fruits, the eggs are seldom laid on healthy undamaged fruits or pods. Some damage to the skin or peel is normally necessary for the penetration of the young larvae and previous damage by other insects, such as the exit holes of carob midges on carobs or Bruchidae on *Acacia* pods has been shown to encourage *Ectomyelois* infestation. Eggs are also laid in the cracks which occur naturally in the skins of carob pods. Dampness also favours *Ectomyelois* by encouraging the growth of moulds and other micro-organisms which render the fruit more suitable for the larvae. On almonds, infestation depends on the presence of some nuts with cracked shells and a further generation of moths may be produced in the dried product during storage. (Gothilf, 1970; Calderon, Navarro and Donahaye, 1969; Avidov and Gothilf, 1960).

In the laboratory, no development was achieved at 15°C. On ripe carobs egg incubation ranged from eight days at 20°C to three days at 30°C and 34°C. Larval development ranged from 45 days at 20°C to 26 days at 30°C and 31 days at 34°C (Gothilf, 1969). Cox (1976) found that adult activity and mating only took place in the laboratory under fluctuating conditions of temperature, humidity and, above all, light, resembling the natural habitat. At 70 per cent relative humidity he recorded average life cycles, from egg to adult, of 48 days at 20°C, 30 days at 25°C and 23 days at 30°C. In laboratory experiments larvae entered diapause in response to low temperatures and short photoperiods (Cox, 1979).

Ephestia calidella Guenée – – carob moth, dried fruit moth

(= *Cadra calidella* Guenée)

Distribution: Mediterranean region.

Interceptions:

? Chile:	prunes
Algeria:	carobs
Crete:	carobs
Cyprus	carobs, dried fruit, almonds, vetches
Gibraltar:	almonds
Greece:	carobs, dried fruit
Iran:	almonds
Iraq:	dates
Italy:	almonds, hazelnuts
Lebanon:	dates, lentils, groundnuts
Morocco:	carobs
Portugal:	carobs, dried fruit, almonds
Spain:	dried fruit, carobs
Syria:	gallnuts
Turkey:	carobs, dried fruit, hazelnuts
Kenya:	oilcake
Tanzania:	oilcake

During the 21 years, 1957 to 1977, this species was recorded on 365 occasions from Mediterranean cargoes, of which 258 were of carobs, 60 of dried fruit, 36 of nuts and 11 of other commodities. There were single records from Chile in 1969 and from Kenya and Tanzania in 1961.

Biology: this species is known best as a pest of dried fruits and carobs and is able to attack these crops in the fields before harvest as well as the dried product in stores. It is, therefore, both a field and a storage pest.

Its temperature and humidity requirements are similar to those of *Ephestia cautella*, although it is much more restricted in its food preferences. In laboratory studies, development has been achieved at temperatures ranging from 15°C to 35°C, although mortality is very high at 15°C and only a very few individuals were able to complete their life cycle at that temperature. The shortest life cycle (23 days) was achieved at 30°C and 70 per cent relative humidity. At lower humidities, development is retarded but is still possible as low as 20 per cent.

At temperatures below 25°C, the fully grown larvae may enter diapause which can prolong the larval period to over 500 days in some individuals. Studies of the diapause suggest that its onset is influenced by the shortening of the day length as well as by the lowering of temperature, and in nature it is the normal reaction to the advance of autumn into winter. The same sort of diapause is also to be found in *Ephestia figulilella*, *E. elutella* and *Plodia interpunctella*, and to a lesser degree in *E. cautella* and *E. kuehniella*. The diapause confers several advantages — it enables the species to survive periods of adverse climatic conditions and provides a synchronisation of adult emergence as soon as favourable conditions return. In addition, diapausing larvae are more resistant to insecticides than feeding larvae. The diapause, therefore adds considerably to the potential of these species as stored products pests, especially in temperate climates. (Prevett, 1968; Omar, Kamal, El-Kifl and Wahab, 1974; Cox, 1974, 1975a and b).

Ephestia cautella (Walker) – – tropical warehouse moth

(= *Cadra cautella* (Walker))

Distribution: very abundant in warm and tropical countries throughout the world. It is very frequently imported into Britain where breeding takes place in warehouses in the summer although mortality may be very heavy in winter (Burges and Haskins, 1964).

Interceptions: (see Table 14)

Table 14

Percentage of cargoes found infested with Ephestia cautella *(Walker) where 10 or more cargoes were inspected.*

	Nearctic	Neotropical	Ethiopian	Northern Europe	Mediterranean	Oriental	Australasian	Total nos. of cargoes
Grain	2	21	24	<1	<1	25	9	2048
Grain products	5	52	39	<1	10	42	5	2828
Oilseeds	8	26	50	<1	14	37	23	4422
Oilcake	7	42	58	0	10	50	7	7971
Copra	0	0	61	0	0	17	55	201
Cocoa beans	0	68	63	0	0	25	71	2349
Dried fruit	<1	6	3	1	13	19	2	1054
Nuts	<1	16	8	<1	9	22	7	1049
Carobs and carob meal	0	0	0	0	50	0	0	1068
Spices	5	13	12	0	2	6	0	486
Pulse and pulse products	<1	6	13	0	4	17	<1	587
Animal products	0	4	5	0	<1	4	<1	326
Others	<1	5	7	<1	3	7	6	1082
Total nos. of cargoes	509	3244	10483	23	2894	7225	1093	25471

Throughout the 21 year period analysed here, this species was the second most frequently recorded insect on imported cargoes, outnumbered only by the tenebrionid beetle *Tribolium castaneum* (Herbst). Infestations were particularly frequent from all tropical regions of Africa, Asia, South America and the West Indies and on a very wide range of vegetable produce.

As with most other common insect pests, *E. cautella* showed some decrease in its frequency although on many commodities this was relatively slight. The incidence of infestation tended to fluctuate widely from year to year and on many commodities a decline in infestation levels is only apparent when an over-all comparison is made between the first ten years and the last eleven years of the period.

In this way, a decrease in infestation is seen from Argentinian grain products and oilcake, with 58 per cent and 60 per cent respectively of cargoes affected in the first part of the period, falling to 10 per cent and five per cent in the last part. Infestation of dried fruit from the Mediterranean region also showed a marked decline, due mainly to a drop in infestation levels on Greek dried fruit, from 34 per cent of cargoes in the first part of the period to four per cent in the last part.

On the other hand, some commodities showed little or no decline at all in infestation. Frequency of infestation of cocoa beans from the West Indies and the Pacific islands, for instance, remained consistently high — 70 per cent or more of cargoes in both parts of the period. In spite of the marked decline on Mediterranean dried fruit, Mediterranean carobs and carob meal showed a lesser decrease, from 55 per cent to 40 per cent of cargoes.

From tropical Africa and Asia a few commodities showed a decline in infestation. Ghanaian cocoa, for instance, showed a decrease from 60 per cent to 42 per cent of cargoes in the two parts of the period, Kenya oilcake from 63 per cent to 42 per cent, Indian oilcake from 53 per cent to 37 per cent and Burmese oilcake from 76 per cent to 54 per cent. Such decreases, however, are far from universal and in a great many cases the frequency of infestation showed either no decrease at all or, at best, only a slight one.

Biology: development from egg to adult emergence can take place within the range 15°C to 36°C and at relative humidities in excess of 20°C. The length of life cycle can range from 194 days at 15°C to 34 — 41 days at 35°C. The shortest life cycle (29 to 30 days) was achieved at 30 to 32°C and 70 to 80 per cent relative humidity.

Diapause in this species was first definitely identified by Hagstrum and Sharp (1975) in a population infesting a warehouse in Florida. Since then the effects of temperature and photoperiod on diapause in populations from different parts of the world have been studied by Bell and Bowley (1980) and Bell, Cox, Allen *et al* (In press). They found that most of the populations tested had some tendency to diapause, although usually only a proportion of larvae could be induced to diapause under any of the conditions tested. The duration of diapause was generally quite short at 20°C, mean time to pupation ranging from about four to 15 weeks. Many individuals showed a very transient diapause lasting only a few weeks. Diapause is thus unlikely to enable *E. cautella* to overwinter in Britain in unheated premises, although larvae can survive within stacks of bagged commodities or bulks of grain which have been in store during the summer and which retain some warmth.

Ephestia elutella (Hübner) – – warehouse moth

Distribution: in most temperate regions of the world, where it is found mainly in the protected environment of warehouses, factories etc. and hardly ever on farms or in the open. Absent from the tropics.

Interceptions: (see Table 15)

Table 15

Records of Ephestia elutella (*Hübner*)

Total number of records = 142

Commodity group	Number of records	Geographic region	Number of records
Grain	54	Nearctic	5
Grain products	26	Neotropical	7
Oilseeds	2	Ethiopian	19
Oilcake	3	Northern Europe	4
Cocoa beans	6	Mediterranean	37
Dried fruit	29	Oriental	2
Nuts	7	Australasian	68
Carobs	4		
Others	11		

Despite its great abundance as a warehouse pest in Britain and other temperate countries, this species is only infrequently imported on food cargoes. About half of the recorded imports were from Australia, mainly on grains and flour. Mediterranean cargoes, particularly dried fruits were the next most common source of infestation. Ten of the records from the Ethiopian region were from South Africa, mainly from maize and maize products.

There are, however, a few records from tropical countries which would seem to be very unlikely sources of this essentially temperate species. It was, for instance recorded four times from Nigerian cargoes, in spite of the fact that Howe (1952) states that it has not been found in any part of Nigeria. It was also recorded once from the Ivory Coast, once from Burma and twice from the West Indies. One cannot be certain of the origin of these specimens.

Biology: this species is one of the most serious pests in warehouses, stores and factories in all temperate regions of the world. The following information is taken from Richards and Waloff (1946), Waloff (1948 and 1949), Strumpel (1964), Bell and Walker (1973) and Bell (1975 and 1976).

The adult moths begin to emerge in late spring and emergence continues throughout the summer. The larvae will feed on a very wide range of raw materials and finished products, including cocoa beans and manufactured chocolate, dried fruit, nuts, oilseeds and oilcakes, grains, grain products and tobacco. In the autumn, at the end of their feeding cycle, the fully grown larvae come to the surface of the stack or bulk of food, where they wander about, spinning fine threads of silk as they go. A heavy infestation will produce such a quantity of webbing as to cover the surface of the stack with a continuous sheet. After a few days, the wandering larvae migrate away from the food to spin their cocoons in cracks and crevices of the building structure. A small proportion of the larvae pupate right away to give a small second generation of adults but most of the larvae enter diapause in which they spend the winter.

As one might expect from its temperate distribution, *E. elutella* breeds within a lower temperature range than its tropical counterpart *E. cautella*. The lower limit for development is just below 15°C and at the upper end of its range, development is just possible at 30°C although adults produced at this temperature are infertile. As the environmental temperature falls, the proportion of larvae entering diapause increases. The shortening of the day length as the season advances from autumn to winter also exerts an important influence on the onset of the diapause.

Its characteristic life history and in particular its diapause makes *E. elutella* a highly successful and damaging pest. In their diapaused state the larvae are very resistant to cold and are well protected against insecticides. Moreover, the diapause usually lasts for eight or nine months, during which time a complete turn-over of warehouse stock may take place without having any effect on the endemic population of this pest. Not only do the larvae damage goods by feeding directly upon them, they also spoil more than they eat by fouling it with their copious webbing.

Ephestia figulilella Gregson – – raisin moth, fig moth

Distribution: Mediterranean region, also in those parts of North America and Australia which have a Mediterranean-type climate.

Interceptions:

Crete:	carobs, dried fruit
Cyprus:	carobs and carob meal, almonds, dried fruit
Greece:	dried fruit
Iran:	dates
Iraq:	dates
Lebanon:	groundnuts
Portugal:	dried fruit
Spain:	dried fruit, almonds
Turkey:	carobs, dried fruit
Australia:	dried fruit

During the 21 years 1957 to 1977, this species was recorded 113 times from imports. Of these, 33 were from carobs and carob meal and 75 from dried fruit.

Biology: Ephestia figulilella has similar food preferences to *E. calidella* and is recorded mainly from dried fruit and carobs. It is not, however, commonly found in stores and is more frequently associated with fallen fruit and freshly harvested carob and fruit crops. It is also recorded as a serious pest of ripening dates on the trees.

Its temperature and humidity requirements are similar to those of *E. cautella*, although it is much more restricted in its food preferences. In the laboratory, development has been achieved within the range 15 to 35°C. The shortest average life cycle (27 days) occurs at 30°C and 70 to 90 per cent relative humidity. At the cooler end of its range, the fully grown larvae enter diapause. Studies of the diapause show that its onset is governed by the shortening of the day length as well as by a fall in temperature. The same sort of diapause is found in *E. calidella, E. elutella, Plodia interpunctella*, and to a lesser degree in *E. cautella* and *E. kuehniella* (Cox, 1974; 1975a and b).

Ephestia kuehniella Zeller – – Mediterranean flour moth, mill moth

(= *Ephestia sericarium* Scott)

(= *Anagasta kuehniella* (Zeller))

Distribution: throughout the temperate regions of the world, absent from the tropics. Freeman (1962) in a survey of flour mills in various parts of the world, found that this species is dominant in mills in 'moderate climates' where the average temperature is below 22°C and the relative humidity over 55 per cent. In hot humid regions of the world, *E. kuehniella* is replaced by *Corcyra cephalonica* as a mill pest. Solomon and Adamson (1955) found that it could withstand rigorous winter conditions in Britain although it does not thrive in unheated buildings, preferring the warmer conditions of flour mills and bakeries.

Interceptions: (see Table 16)

Table 16

Records of Ephestia kuehniella *Zeller*

Total number of records = 493

Commodity group	Number of records	Geographic region	Number of records
Grain	58	Nearctic	19
Grain products	384	Neotropical	125
Oilseeds	8	Ethiopian	200
Oilcake	10	Northern Europe	27
Pulse and pulse products	15	Mediterranean	34
Others	18	Oriental	3
		Australasian	85

During the 21 year period there has been a marked decline in the number of records, from 139 in 1957 to only 1 in 1977. From the Neotropical region the majority of records (121) were from Argentinian cargoes and all of these occurred during the early part of the period, between 1957 and 1964. Argentinian wheat products were the most commonly affected cargoes and during these years the species was found on about 12 per cent of them. Other South American records were two from Uruguay rice, one from Chilean prunes and one from Mexican maize.

From the Ethiopian region, 176 of the 200 records were from South Africa, again during the first part of the period. South African maize products were the most commonly affected cargoes and during the eight years 1957 to 1964 about 13 per cent of these cargoes carried the species. Other African records were from Mozambique (16), Tanzania (three) Kenya and Rhodesia (two each) and Sudan (one). There were no records from West African cargoes during the period analysed here, although Howe and Freeman (1955) had recorded it earlier on West African imports. They suggested, however, that these specimens came from infested residues in the ships' holds since the species is probably unable to breed in the West African climate.

Most of the Australian records were from cargoes of wheat products. Although the species occurred on imports from Australia fairly consistently throughout the period, it affected only a small proportion of wheat products cargoes — never more than seven per cent in any year and generally less than five per cent.

Biology: this species is best known as a pest of flour mills. The larvae feeding in the flour construct tubes of silk, in which they live. A heavy infestation can seriously affect the working of a flour mill because the flour becomes felted by the silk strands and webbed into lumps, clogging the machinery and blocking spouts. Infestations may build up to such proportions as to stop production and necessitate arduous and expensive cleaning.

As one might expect from its temperate distribution, *E. kuehniella* breeds within a temperature range lower than *E. cautella* but similar to that of *E. elutella*. The lower limit is about 12°C and breeding is just possible at 30°C although the adults produced at this temperature are infertile. Jacob and Cox (1977) reared the larvae on white flour, a poor food but one that is often infested. They found that optimum conditions were at 25°C and 75 per cent relative humidity at which the life cycle from egg to adults took an average of 74 days. Other authors (Hassanein and Kamel, 1965; Siddiqui and Barlow, 1973; Bell, 1975) using a variety of foods, obtained life cycles ranging from 143 days at 15°C to as little as 33 days at 30°C. Maize products were found to be a more satisfactory diet than either wheat or rice products. During a population study of moths in a large unheated grain store in Scotland, some *E. kuehniella* larvae survived the winter by remaining as full-fed last instar larvae from the autumn until the following May before pupating (Cole and Cox, 1981). Laboratory studies confirmed that these larvae had entered diapause, the incidence of which was increased by rearing larvae in continuous darkness as well as by low temperatures (Cox *et al.*, 1981b).

Ephestia sericarium Scott, see *Ephestia kuehniella* Zeller

Mussidia nigrivenella Ragonot

Distribution: tropical Africa. Howe and Freeman (1955) record it on imports from West Africa, Forsyth (1966) records it in Ghana, Le Pelley (1959) in Kenya and Uganda, and Food and Agriculture Organisation (1973) in Tanzania.

Interceptions:

Ghana:	cocoa beans
Nigeria:	butter beans, sheanuts
Sierra Leone:	calabar beans
West Africa:	butter beans, palm kernels

This species is only infrequently found on imported cargoes. According to Howe and Freeman (1955) it was found 10 times on West African cargoes during the period 1942 to 1952 and Hurlock (1959) records it 13 times from 1953 to 1958, from Ghana and Sierra Leone, mainly on calabar beans. Since then it has been rare on imports, recorded only three times — on Nigerian sheanuts in 1959 and on Sierra Leone calabar beans in 1964 and again in 1966.

Myelois ceratoniae (Zeller), see *Ectomyelois ceratoniae* (Zeller)

Plodia interpunctella (Hübner) – – Indian meal moth

Distribution: although this species thrives best in warm climates and is abundant throughout the tropical and subtropical parts of the world, it is also established in the sheltered conditions of factories and stores in temperate zones. Solomon and Adamson (1955) found that it survived severe winter conditions in stores in Britain.

Interceptions: (see Table 17)

Table 17

Percentage of cargoes found infested with Plodia interpunctella *(Hübner) where 10 or more cargoes were inspected.*

	Nearctic	Neotropical	Ethiopian	Northern Europe	Mediterranean	Oriental	Australasian	Total nos. of cargoes
Grain	1	7	7	<1	1	<1	1	541
Grain products	<1	7	8	0	2	<1	<1	328
Oilseeds	3	5	2	<1	5	1	3	224
Oilcake	<1	4	<1	0	0	<1	0	145
Cocoa beans	0	<1	<1	0	0	0	<1	17
Dried fruit	<1	4	<1	0	4	12	2	385
Nuts	1	0	4	6	10	7	9	806
Carobs and meal	0	0	0	0	2	0	0	32
Spices	0	<1	<1	0	2	<1	0	21
Pulses and pulse products	<1	1	<1	<1	<1	1	<1	45
Others	<1	<1	<1	<1	1	<1	0	65
Total nos. of cargoes	225	393	496	80	1002	265	118	2609

During the 21 year period, there has been a decline in the number of records of the species, from 335 in 1957 and 372 in 1958 to only 45 in 1977.

From the Neotropical region, Argentinian grain and grain products were the most frequent sources of the species, with the greatest frequency occuring in 1960. In that year, 27 per cent of Argentinian grain and 15 per cent of grain products cargoes were affected but by 1977 only six per cent of grain cargoes carried the species.

From the Ethiopian region, South African grain and grain products were commonly affected commodities although, again, there has been a considerable decline in the proportion of cargoes infested, from 12 per cent of grain and 16 per cent of grain products cargoes in 1957 to none on either commodity by 1977. Howe and Freeman (1955) reported that *Plodia* was particularly associated with West African groundnuts, affecting 32 per cent of groundnut kernel cargoes between 1942 and 1946 and 34 per cent between 1947 and 1952. Towards the end of this latter period, however, a considerable drop in infestation levels occurred, from 50 per cent in 1947 to less than 20 per cent of cargoes in 1951 and 1952. This fall in infestation of West African groundnuts has been continued during the period of the present survey. In 1957, 18 per cent of cargoes were affected and the level fell steadily to three per cent in 1973, rising to six per cent in 1975 and falling to nil in 1977.

From the Mediterranean region, most of the records were from dried fruit and nut cargoes, with a decline in the proportion of affected cargoes, from 17 per cent of dried fruit and 16 per cent of nuts in 1957 to two per cent and none, respectively, in 1977.

The species was very common from northern Europe and most of the records were from shipments of French walnuts, again showing a decrease from 21 per cent of cargoes in 1957 to none in 1977.

Biology: in the following account, I have drawn on the works of Tzanakakis (1959), Hassan, Hassanein and Kamel (1962), Tsuji (1963), Williams (1964), Morere and Le Berre (1967), Abdel-Rahman (1971), Savov (1973) and Bell (1975 and 1976).

The larvae feed on a wide variety of vegetable material but they are particularly noted as pests of grain (of which they consume the germ only), dried fruit and nuts. Damage is caused not only by the direct feeding of the larvae but also by the spoilage of the food by frass and webbing rendering it unfit for human consumption. The number of generations produced in one year may vary from one to two in Britain and Europe to as many as eight in warm climates such as Egypt and California. The winter is spent in a state of diapause in which the larvae are able to survive lengthy periods of adverse conditions.

Savov (1973) records a lower temperature threshold of 13.5°C although other authors found that complete development was impossible at 15 or 16°C. The mean length of larval development ranged from 112 days at 20°C to 34 days at 30°C and 27 days at 35°C. Although the quickest development occurred at 35°C, the optimum for survival was 30°C. Studies of the diapause reveal that several factors may be responsible for its induction — crowded conditions in early larval life, a progressive fall in temperature during development and a shortening of the day length all play a part in producing diapause in the autumn generation of larvae.

Spectrobates ceratoniae (Zeller), see *Ectomyelois ceratoniae* (Zeller)

Pyralinae

This sub-family is best known for the species *Pyralis farinalis* (L.), which occurs as a minor pest of flour mills and is the largest and most conspicuously marked of all the stored products moths. Some other species listed here are recorded from time to time on stored products, but none of the Pyralinae approach the Phycitinae in importance as stored products pests.

Pyralis farinalis (L.) – – meal moth

Distribution: cosmopolitan, but more plentiful in temperate than in tropical regions (Corbet and Tams, 1943). McFarlane (1963) records it in a feed mill in Jamaica, where it was not common. Howe and Freeman (1955) record it on West African imports but suggest that it is incapable of living in the tropics and that these imported specimens were probably acquired from endemic populations in the ships' holds. Ruppel (1977) considered it to be the most serious pest in grain stores of Michigan, U.S.A.

Interceptions: (see Table 18)

Table 18

Records of Pyralis farinalis (*L.*)

Total number of records = 53

Commodity group	Number of records	Geographic region	Number of records
Grain	19	Nearctic	7
Grain products	9	Neotropical	23
Oilseeds	3	Ethiopian	4
Oilcake	6	Northern Europe	6
Dried fruit	5	Mediterranean	8
Pulse and pulse products	4	Oriental	2
Animal products	3	Australasian	3
Others	4		

From South America, 19 of the records were from Argentina. Other South American records were from Uruguay rice, Trinidad cocoa and Brazilian Brazil nuts and rice. From Africa, there were single records from Tanzania and Kenya and two from Nigeria. From the Orient, the species was recorded only from China, on rice and gallnuts.

Hurlock (1963, 1964) lists *Pyralis farinalis* as a species that is probably associated with moulds and residues in the ships' holds and it seems likely that some, at least, of the infestations recorded here were acquired from that source and not directly associated with the cargoes on which they were found. This is probably the case, in particular, with specimens found on cargoes from hot countries such as Nigeria, Brazil and Trinidad.

Biology: the species is well known as an inhabitant of flour mills, where it is usually only a minor pest, feeding mainly in accumulated residues where hygiene is poor. Woodroffe (1953) records it as sometimes abundant in pigeon nests in Britain and its occasional appearance in domestic and storage premises may be due to its spread from nests in the building structure.

Pyralis manihotalis Guenée

Distribution: Indo-Australian region, tropical and South Africa, tropical America, Hawaii (Corbet and Tams, 1943). Forsyth (1966) records it on stored cocoa and maize in Ghana, and Cornes (1973) records it on cassava flour and groundnuts in Nigeria.

Interceptions:

Brazil:	Brazil nuts
Kenya:	cottonseed cake, coconut shell
Nigeria:	kola nuts, ginger, bones and horns, hooves horns and timber
Uganda:	coffe beans
Pakistan:	walnut kernels

Tineidae

Among the moths of this family are species whose larvae feed mainly or wholly on materials of animal origin. Some of these species are common inhabitants of bird and rodent nests, feeding on fur, feathers and the dead remains of other insects, and such nests probably represent the original natural habitats of these species. It is a short step from infesting the nests of birds and rodents to infesting the homes of human beings, and some of the best known tineids are household pests such as the common clothes moth (*Tineola bisselliella*) and the case bearing clothes moth (*Tinea pellionella*) which are capable of causing serious damage to clothes, woollen goods, furnishings, pipe laggings etc. in dwelling houses.

The dietary requirements of many Tineidae are not fully known but is would appear that by no means all of them are restricted to animal materials. Several species are found in regular association with stored food products. Vegetable matter, particularly moulds, almost certainly forms an important part of their diet. Some are to be found as pests in wine cellars, feeding on moulds and attacking the wine corks, causing serious losses of wine through leakage and spoilage.

Acedes fuscipunctella (Haworth), see *Niditinea fuscipunctella* (Haworth)

Acedes pallescentella (Stainton), see *Tinea pallescentella* (Stainton)

Nemapogon granella (L.) – – corn moth, European grain moth

(= *Tinea granella* (L.))

> *Distribution:* nearly cosmopolitan, in the United Kingdom, most parts of Europe, Japan, Formosa, U.S.A., Argentina (Hinton, 1956). Howe and Freeman (1955) record it from Est African imports and suggest that, although it has not been recorded in West Africa, it could probably live there.

> *Interceptions:* (See Table 19)

Table 19

Records of Nemapogon granella *L.*

Total number of records = 53

Commodity group	Number of records	Geographic region	Number of records
Grain	18	Nearctic	9
Grain products	7	Neotropical	5
Oilcake	3	Ethiopian	4
Nuts	16	Northern Europe	31
Animal products	0	Mediterranean	2
Others	9	Oriental	0
		Australasian	2

This species, and *Niditinea fuscipunctella*, are the two most common tineids found in association with stored products and the only two to be recorded with any regularity on food cargoes. The Ministry's records suggest that the species has a mainly temperate range, occurring most often on North American and European cargoes. Fifteen of the records from nuts were from cargoes of French walnuts, particularly during 1957 and 1958.

The Ministry also holds numerous records from residues in ships' holds.

Hurlock (1963, 1964) lists this species as one that lives on mouldy and damp cargo residues and some of the infestations found on cargoes probably came from such residues and were not necessarily associated with the cargoes on which they were found.

Biology: according to Hinton (1956), it is common out of doors in fungus and is known as a pest of grain and other stored cereals, particularly when the moisture content is high. The larvae may sometimes bore into the decayed wooden structure of warehouses, mills and other buildings. It is also known as a pest in wine cellars. Bender (1941) studied the species in wine cellars in Germany where the larvae attack wine corks and feed on the parts of the corks overgrown with fungus. Zagulyaev (1966) records it amongst the Lepidoptera which prey in their larval stage on the lac insect *Laccifer lacca* (Kerr) in the USSR.

Niditinea fuscipunctella (Haworth) – – brown-dotted clothes moth

(= *Acedes fuscipunctella* (Haworth))

(= *Tinea fuscipunctella* (Haworth))

Distribution: cosmopolitan. Cornes (1973) records it in stored groundnuts in Nigeria. Legner and Eastwood (1973) describe it as a pest of households and poultry farms in California.

Interceptions: (see Table 20)

Table 20

Records of Niditinea fuscipunctella (Haworth)

Total number of records = 54

Commodity group	Number of records	Geographic region	Number of records
Grain	21	Nearctic	11
Grain products	4	Neotropical	13
Oilcake	8	Ethiopian	11
Cocoa beans	2	Northern Europe	4
Dried fruit	3	Mediterranean	6
Nuts	1	Oriental	7
Animal products	7	Australasian	2
Others	8		

In addition to these records from imports, the Ministry holds a great many records from cargo residues in ships' holds and Hurlock (1963 and 1964) lists it as a species which lives in mouldy and damp cargo residues. It seems likely that some of the infestation found on cargoes came from such residues and not necessarily from the cargoes on which they were found.

Biology: according to Hinton (1956), this species is recorded on grain and other stored products but the larvae are unable to develop on whole grains alone and may feed on mite excrement and other refuse. In wheat they are often associated with mites. In the laboratory the larvae feed readily on hair, flannel, soiled feathers etc. Woodroofe (1953) records it twice from birds' nests in Britain. Legner and Eastwood (1973) report it breeding in poultry droppings in poultry farms and spreading as a household pest, often in large numbers, to nearby dwellings. Zagulyaev (1966) records it amongst the Lepidoptera that prey, in their larval stages, on lac insects, *Laccifer lacca* (Kerr) in the USSR.

Setomorpha rutella Zeller

Distribution: in tropical and sub-tropical regions of the Old and New Worlds (M.R. Sattler, identifier's comment). Hinton (1956) cites records from the East Indies, India and USSR, Africa, Malaya, Puerto Rico and Java. Le Pelley (1959) records it in Tanzania, Forsyth (1966) in Ghana and Cornes (1973) in Nigeria.

Interceptions:

Brazil:	Brazil nuts
Sierra Leone:	calabar beans
South Africa:	maize
Bangladesh:	dried ginger, crushed bones
Burma:	cheroots
India:	rice bran, dried ginger
Samoa:	cocoa beans

This species is known as a pest of stored dried tobacco but cargoes of tobacco are not regularly examined by the Ministry's Inspectorate and I am unable to give any data for importations on that commodity.

Biology: apart from stored dried tobacco, with which this species is particularly associated, it is recorded from a wide variety of materials of both vegetable and animal origin. It has, for instance, been found on stored cereals and seeds, coffee, cocoa, oilseeds and bulbs, animal fibres, skins, bees' and wasps' nests and bird guano (Corbet and Tams, 1943; Hinton, 1956; Le Pelley, 1959; Forsyth, 1966).

Tinea fuscipunctella (Haworth), see *Niditinea fuscipunctella* (Haworth)

Tinea granella (L.), see *Nemapogon granella* (L.)

Tinea pallescentella Stainton – – large pale clothes moth

(= *Acedes pallescentella* (Stainton))

Distribution: Europe, introduced into North America (Hinton, 1956). McFarlane (1963) records it as common and widespread in Jamaica. The Ministry's records suggest a widespread, possibly worldwide distribution.

Interceptions:

Canada:	wheat
United States:	mohair, oilcake
Argentina:	casein, hooves and horns
Faroe Islands:	fishmeal
Nigeria:	dried blood
South Africa:	maize
Tanzania:	cottonseed husks

Denmark:	fishmeal
Finland:	wheat
France:	maize
Holland:	feathermeal
Sweden:	dried skins
Italy:	cornflour
Pakistan:	wool
New Zealand:	dried blood

Specimens recorded from vegetable materials, such as oilcake, wheat and maize may represent cross-infestations from animal products cargoes or from endemic populations feeding on residues in the ships' holds.

Biology: the larvae of this species feed only on materials of animal origin. Hinton (1956) records it in stables, outhouses, mills, granaries, warehouses and houses, on woollen materials, hair, skins, dead animals and in bird and wasp nests. Woodroffe (1953) records it twice in pigeon nests, occurring in considerable numbers.

Tinea pellionella (L.) – – case-bearing clothes moth

Distribution: Temperate and cool Mediterranean zones in the Palaeartic and Nearctic regions and introduced into Australia and New Zealand (Robinson, 1979). Le Pelley (1959) records it in Kenya and Uganda. Howe and Freeman (1955) record it on West African imports but suggest that it is incapable of living in the tropics and that such infestations probably came from residues in the ships' holds.

Interceptions:

Canada:	wheat, soya bean meal
United States:	mohair, soya bean meal
Argentina:	rice, dried offal
Nigeria:	hooves and horns
Portugal:	bone meal
Malaya:	hooves

Biology: the larvae feed on substances containing keratin, such as wool, hair, fur and feathers. Each larva constructs a tubular case, open at both ends, which it makes out of fibres cut from the food material bound together with silk. The larva is able to turn around inside the tube and feed from either end, and it is also able to move around on the surface of the material, carrying the case along with it.

It is a widespread household pest, capable of doing damage to clothes and woollens, particularly soiled or unwashed wool, feathers in pillows and horsehair padding in furniture. It is however, of considerably less importance than the common clothes moth, *Tineola bisselliella*. In Britain at least, domestic outbreaks of *pellionella* are now rare and apparently restricted to houses which are damp, poorly heated or unoccupied. (Robinson, 1979).

Woodroffe (1953) records it as abundant and widespread in British birds' nests, especially those of jackdaws and pigeons.

Tineola bisselliella (Hummel) – – common cloths moth

Distribution: widespread throughout the world. According to Howe and Freeman (1955) it is unable to live in the tropics and records of this species on West African imports probably represent cross-infestations from residues in the holds of ships. The Ministry also holds records of this species collected by R.H. Thompson from felt packing in Cyprus.

Interceptions:

Canada:	feather meal, mohair, bone meal, dried blood meal
United States:	mohair, hides
South Africa:	sheepskins
Russia:	camel hides
Australia:	hides, blood meal

It has also been found on whalemeat meal in a whaling factory ship.

Biology: this is the common clothes moth which has been one of the most serious household pests in Britain, doing considerable damage to clothing and furnishings. In 1948, damage by this species was put at £1½ million although it has greatly decreased in importance since then. Its place as a household pest is being taken by carpet beetles (*Anthrenus* spp.) and fur beetles (*Attagenus* spp.) (Hickin, 1974). In its natural habitat, it probably maintains itself on insect remains, fur and feathers, although it seems doubtful whether it is able fully to exploit birds' nests as a habitat and in Britain at least, it is not a regular nest-dwelling species. (Woodroffe, 1953).

The larvae feed, in the main, on materials of animal origin, producing copious silk which covers the food material as webbing. They are also carnivorous to some extent, feeding on other insects and mites. They will, however, feed on vegetable materials and occasional infestations have been found on stored grains, grain products and other vegetable produce. Mr G.A. Brett, in a private communication, records finding the species feeding on flour stored in a former woollen clothing warehouse. Development is possible within the range 10 to 33°C. Shortest development (39 days) is at 25°C on a diet of fishmeal. (Notini, 1939; Geigy and Zinkernagel, 1941; Griswold, 1944; Hinton, 1956; Hickin, 1974).

Diptera

These are the true flies, in which only the front pair of wings are functional. The hind wings are modified into halteres, which act as gyroscopic stabilisers during flight. They are a very diverse order and have adapted themselves to live in almost every ecological niche, even the sea. This diversity, coupled with the aerial mobility of the adults has resulted in their frequent appearance as strays on cargoes, away from their normal habitat.

Several classifications of the Diptera have been proposed (Oldroyd, 1964, 1977; Colyer and Hammond, 1968) but I have chosen to use here the one that is most widely used, in which the group is divided into three sub-orders the Nematocera, Brachycera and Cylorrhapha. In the Nematocera are grouped the most primitive flies, including the crane flies, mosquitoes and those often fragile species that are usually referred to as gnats and midges. They have simple, often elongated, antennae and are usually weak in flight. The Brachycera are intermediate between the primitive Nematocera and the advanced Cyclorrhapha, tending to be more robust in their bodies than the Nematocera and with some reduction in their antennae. The Cyclorrhapha are the most advanced of the Diptera and include the house flies and blow flies as their best known examples. They have much reduced antennae, robust bodies and, with few exceptions, notably amongst the Hippoboscidae, are capable of strong flight.

Brachycera

Scenopinidae – – window flies

The biology of these flies is little known but several species are known to have predaceous larvae which prey on the larvae of Dermestidae in stores and museums. They may also occur in carpets, where they are predaceous on the larvae of moths.

Scenopinus fenestralis L.

   Distribution: world-wide.

   Interceptions:
      Senegal: groundnut cake
      Greece: currants, carobs
      Lebanon: bone meal

Biology: the larvae of this species are predaceous on clothes moth, house moth and flea larvae and are favoured by dry dusty habitats (Colyer and Hammond, 1968). Woodroffe (1953) records the larvae as second only to *Lyctocoris* (Hemiptera) as predators in birds' nests.

Scenopinus glabrifrons (Meigen)

   Distribution: world-wide

   Interception:
      Crete: carob meal

Cyclorrhapha.

Drosophilidae

The members of this family are quite varied in their habits but the ones which are of most significance to us are those that are associated with fermenting substances and which are commonly called fruit flies or vinegar flies. They are attracted, in particular, to the odours of over-ripe and fermenting fruit, sap exuding from trees and decaying fungi.

Drosophila hydei Sturtevant

   Distribution: cosmopolitan

   Interception:
      Peru: cottonseed meal

Drosophila melanogaster complex

   Distribution: world-wide in tropical and temperate regions.

   Interceptions:
      Peru: cottonseed meal
      Gambia: onions
      Ivory Coast: coconuts
      South Africa: dried citrus fruit peel

West Africa: palm kernels, hold after cargo of logs
Cyprus: 'blown' cans of grapefruit
Greece: currants
Iran: raisins, sultanas, dried apricots
Spain: canned preserves
Turkey: dried fruit
India: celery seed, canned goods
Australia: raisins, currants, sultanas, hides
Samoa: cocoa beans

Biology: these insects are very well known because of their use as experimental insects in genetical and physiological studies, They are rarely found in wild habitats and are usually associated with human habitation and garbage (Wheeler and Takada, 1964). Females may lay from 400 to 900 eggs and the life cycle is completed in 30 days at 15°C, 14 days at 20°C, 10 days at 25°C and 7.5 days at 30°C (Hicken, 1974).

Drosophila polychaeta Patterson and Wheeler

Distribution: Central and South America, Texas, Hawaii, Micronesia (Wheeler and Takada, 1964).

Interceptions:
  Nigeria: cocoa beans
  West Africa: cocoa beans

On deck of cargo ship returing from Malaya

Piophilidae

The larvae of *Piophila* spp. breed in animal material, particularly in preserved meat, ham, bacon, dried fish, cheese and corpses (Oldroyd, 1964). They have often been recorded in vast numbers on wet salted hides, though not on dry hides.

Piophila casei (L.) – – cheese skipper

  Distribution: cosmopolitan

  Interceptions: (*see* Table 21)

Table 21

Records of Piophila casei *(L.)*

Total number of records = 152

Commodity group	Number of records	Geographic region	Number of records
Hides	109	Nearctic	40
Bones and bone meal	14	Neotropical	19
Other animal products	6	Ethiopian	8
Vegetable products	21	Northern Europe	2
Others	2	Mediterranean	22
		Oriental	16
		Australasian	45

Biology: this species is usually associated with high protein material and is often recorded as a pest in provision shops, breeding in scraps of ham, bacon, cheese etc., and in tanneries and fur stores. The larvae are commonly known as cheese skippers because of their ability to leap into the air, up to a height of 7.5 cm. This is achieved by looping the body, grasping the tip of the abdomen with the mouth hooks and suddenly releasing its hold. (Oldroyd, 1964).

Hymenoptera

The Hymenoptera is a very large order of insects which, typically, bear two pairs of membranous wings, coupled together by a series of hooks arising from the front of the hind wing which engage with the down-folded hind margin of the fore wing. Wingless forms occur in many of the families, especially amongst the sterile castes in social Hymenoptera. The order includes a great diversity of insects, ranging from the relatively unspecialised sawflies, in which most of the characters of primitive winged insects may be found, to the extremes of specialisation of structure and behaviour found in the higher parasitic forms and social species.

The Hymenoptera fall naturally into two sub-orders, the Symphyta and the Apocrita.

The Symphyta consists of six super-families containing at least 14 families. Of these, two species of Siricidae (wood wasps or horntails) and one of Tenthredinidae (sawflies) have been found in the course of this work. They are the most primitive and least specialised of the Hymenoptera, there is only a partial amalgamation of the first abdominal segment with the thorax and the abdomen lacks the constricted waist, so characteristic of the Apocrita. The ovipositor is used for egg laying only, modified as a saw or a drill for piercing plant tissue and wood, but never as a sting. The mobile larvae, which are phytophagous, bear thoracic legs and, in the case of the sawflies, abdominal prolegs as well.

The Apocrita is by far the larger sub-order and contains all of the more highly specialised forms. The first abdominal segment is always fused with the thorax and this compound structure is separated from the rest of the abdomen by a constricted waist or petiole, formed by the second abdominal segment. The ovipositor is modified as a sting and the larvae are legless. In most of the Apocrita, the larvae feed on the bodies of other insects or, less commonly, on spiders. Many are internal parasites, developing from eggs that have been inserted through the skin of the living and still active host. Others live as external feeders, developing from eggs that have been laid on the paralysed bodies of stung prey. Many species of ants are fierce predators of other insects while others take a mixed diet of animal and vegetable material. Some Apocrita are entirely vegetable feeders such as the bees, a few of the chalcids that develop in seeds and the Cynipidae that are gall wasps.

The distinction between the parasitic forms and the non-parasitic forms is recognised by the division of the Apocrita into two sections, the Parasitica and the Aculeata. This division, however, is not absolutely clear-cut. The Parasitica, for instance, includes some non-parasitic vegetable feeders such as the seed chalcids and the gall wasps, and the Aculeata, besides containing all of the social nest building species, also includes the parasitic Bethylidae. In spite of this slight overlap between the two sections, I have found it useful to retain this division in the following list of species.

Apocrita — Parasitica

Two of the largest super-families within the Parasitica are the Ichneumonoidea and the Chalcidoidea and all but one of the species listed here fall into one or other of these groups. The Ichneumonoidea are represented by the families Ichneumonidae and Braconidae and the Chalcidoidea by the Chalcididae and Pteromalidae. A third super-family, the Cynipoidea is represented by a single example, *Synergus ruficornis* Hartig in the family Cynipidae.

Most of the Parasitica that are recorded on stored products are parasites of primary stored products pests. A large population of parasites is an indication that the infestation by the host species is of long standing.

Ichneumonidae – – ichneumons, sail wasps

These are ichneumon flies which are parasitic mainly on Lepidoptera and Hymenoptera larvae. Some are large and conspicuous insects. The ovipositor of the female is long and issues far forward beneath the abdomen. The eggs are laid either on the surface of the host's body or beneath the host's skin which is pierced by the ovipositor. The parasitic larvae live within the haemocoel and the host remains alive and feeding until the parasitic larvae are full grown, then the parasites consume the vital organs of the host, leaving the body in order to pupate outside. Some ichneumons are parasitic in woodboring larvae, locating the host larvae within the timber and penetrating the wood with their ovipositors.

Angitia chrysostictos (Gmelin), see *Diadegma chrysostictos* (Gmelin)

Diadegma chrysostictos (Gmelin)

(= *Angitia chrysostictos* (Gmelin))

(= *Nythobia chrysostictos* (Gmelin))

> *Distribution:* widely distributed in Europe and Asia, also probably of wide occurrence in the United States and Canada (Fisher, 1959).

> *Interceptions:*
> Cyprus: carob meal
> Greece: figs

> *Biology:* this is a parasite of many species of Lepidoptera. Fisher (1959) lists 30 lepidopterous hosts and also four species of sawfly hosts. The lepidopterous hosts include the storage species *Ephestia elutella*, *Hofmannophila pseudospretella* and *Endrosis sarcitrella*. In addition it has been bred in the laboratory from *Ephestia kuehniella* and *E. cautella*. Under laboratory conditions it has been bred at temperatures ranging from 15°C to 30°C but it cannot complete development at 10°C or 33°C. At 25°C the life cycle takes 21 to 23 days.

Devorgilla canescens (Gravenhorst), see *Venturia canescens* (Gravenhorst)

Nemeritis canescens (Gravenhorst), see *Venturia canescens* (Gravenhorst)

Nythobia chrysostictos (Gmelin), see *Diadegma chrysostictos* (Gmelin)

Phygadeuon bitinctus (Gmelin), see *Xenolytus bitinctus* (Gmelin)

Venturia canescens (Gravenhorst)

(= *Nemeritis canescens* (Gravenhorst))

(= *Devorgilla canescens* (Gravenhorst)

Distribution: cosmopolitan.

Interceptions:
  Peru: cottonseed meal
  South Africa: kibbled maize
  Cyprus: carobs, carob meal
  Spain: carobs
  Hong Kong: gall nuts

Biology: this is a parasite of lepidopterous larvae. Salt (1976) lists 12 species of Lepidoptera that are natural hosts and a further nine species that have been parasitised in the laboratory. The natural hosts are mainly in the Phycitinae and Tineidae and include the stored products species of *Ephestia* and *Plodia interpunctella*. The parasite larva lives within the hosts' haemocoel and only one larva completes development in each host, irrespective of the number of eggs laid. When fully grown, the parasite usually pupates within the host's cocoon. At 25°C, development takes about 22 days and there are about 16 generations a year. The lower limit for development is 15°C (Kurstak, 1964; Frilli, 1965; Corbet and Rotherham, 1965).

Xenolytus bitinctus (Gmelin)

(= *Phygadeuon bitinctus* Gmelin)

Distribution: cosmopolitan.

Interceptions:
  Argentina: crushed bones
  India: fishmaws

The Ministry also holds numerous records from granaries, mills and warehouses in Britain, usually in association with species of stored products Lepidoptera such as *Hofmannophila pseudospretella*, *Endrosis sarcitrella*, *Ephestia* spp, *Pyralis* and Tineidae.

Braconidae – – braconid wasps

This is a large family, similar in appearance to the Ichneumonidae but distinguishable from them by their generally smaller size and characteristic wing venation. They are parasitic on other insects, most commonly on the larvae of Lepidoptera. Most of them are endoparasites but a few, in the subfamilies Doryctinae and Branconinae, are ectoparasitic (Matthews, 1974).

Allorhogas pallidiceps (Perkins)

Distribution: widespread in the warm regions of the world.

Interception:

　　West Africa: empty hold after discharge of palm kernels and oilcake —
　　　　residual infestation of *Necrobia rufipes* and *Ephestia cautella*
　　　　present

Apanteles carpatus Say

　Distribution: cosmopolitan.

　Interceptions:

　　Western Canada: wheat infested with *Tinea* sp.
　　Peru: fishmeal

Biology: this species is a common parasite of the case-bearing larvae of the
clothes moths *Tineola bisselliella*, *Tinea pellionella* and *Trichophaga tapetzella*.
Woodroffe and Southgate (1951) record it as a frequent inhabitant of birds'
nests in Britain, often occurring in considerable numbers. Fallis (1942)
studied its life cycle as a parasite of *Tineola bisselliella* and found that the total
egg and larval stages ranged from 13 to 45 days at 27°C and from 20 to 154
days at 24°C.

Apanteles trachalus Nixon

　Distribution: Europe.

　Interceptions:

　　France: wheat infested with *Endrosis sarcitrella*, *Hofmannophila pseudos-*
　　　　pretella and *Haplotinea ditella*
　　Italy: prunes, millet sprays

Bracon hebetor Say

(= *Microbracon hebetor* (Say))

　Distribution: throughout the warmer regions of the world.

　Interceptions: (see Table 22)

Table 22

Records of Bracon hebetor *Say*

Total number of records = 660

Commodity group	Number of records	Geographic region	Number of records
Grain	54	Nearctic	7
Grain products	50	Neotropical	92
Oilseeds	57	Ethiopian	180
Oilcake	104	Northern Europe	0
Cocoa beans	119	Mediterranean	219
Carobs and meal	79	Oriental	138
Dried fruit	142	Australasian	24
Nuts	23		
Others	32		

Biology: this species is a very common parasite of stored products Pyralid moths such as *Ephestia* spp. and *Plodia interpunctella*. The females sting and paralyse the late larval stages of the host, laying usually five or six eggs on or beside the immobile victim. On hatching, the parasite larvae attach themselves by their mouthparts to the host, sucking its body fluid while remaining on the surface. At the end of the feeding period, the parasite larvae spin cocoons alongside the shrivelled remains of the hosts. (Soliman, 1942; Morrill, 1942)

Microbracon hebetor (Say), see *Bracon hebetor* Say

Triaspis thoracica (Curtis)

Distribution: Europe, North Africa. Intentionally introduced into the United States, Canada and Australia.

Interception:
Gibraltar: broad bean seed infested with *Bruchus* sp.

Biology: This is a parasite of Bruchidae and has been deliberately introduced into many parts of the world as an agent for the biological control of *Bruchus* spp. on growing pulse crops and seed.

Chalcididae

The Chalcididae belong to the superfamily Chalcidoidea which are commonly known as chalcids. They are small, sometimes minute insects with a very reduced wing venation and geniculate antennae. The great majority of them are parasitic on other insects, but there are a few seed chalcids which develop inside seeds. Three families of chalcids are represented here, the parasitic Chalcididae and Pteromalidae, and the diverse Eurytomidae.

Antrocephalus mahensis Masi, see *Antrocephalus mitys* (Walker)

Antrocephalus mitys (Walker)

(= *Antrocephalus mahensis* Masi)

Distribution: this species seems to extend over most of Africa and has also been taken in south India and Malaya (J.A.J. Clarke, identifier's comment). Howe (1952) and Cornes (1973) record it in Nigeria and Forsyth (1966) records it in Ghana.

Interception:
West Africa: groundnuts

Euchalcidia caryobori Hanna

Distribution: the reference collection of the British Museum (Natural History) records no further distribution than that listed below.

Interceptions:
Sudan: senna pods
Indonesia: illipenuts
Sarawak: illipenuts

Biology: Ferriere and Kerrich (1958) record this species as a parasite of the beetle *Pachymerus* (= *Caryoborus*) *pallidus* Oliv., imported with senna pods. Mr. G.A. Brett, in a private communication, doubts whether the name of the beetle is correct — 'they might mean *Caryedon sudanensis* Southgate, which was not described as a separate species until 1971' (Southgate, 1971).

Eurytomidae

Probably no other family of chalcids exhibits so wide a diversity of habitats. Species are known from galls on wheat, rye, barley and various grasses, orchids, clover and alfalfa seeds, living in nests of bees and wasps, and as parasites of gall forming Diptera and Hymenoptera. A few are egg parasites of Orthoptera. (Imms, 1957)

Systole coriandri Nikol 'skaya

Distribution: southern Europe, Russia, Rumania (reference collection, British Museum (Natural History)). Dr. Bouček, in a personal communication, states that he sees no reason why the specimens recorded here should not be *S. coriandri* as this species is more widely distributed than it was at one time thought.

Interception:

Kenya: coriander seed

Biology: a serious pest of coriander, greatly decreasing the yield of oils and reducing the germination power of the seeds (Ostrovskii, 1940).

Pteromalidae

This, the largest family of chalcids, affects almost all orders of insects, either as parasites or hyperparasites (Imms, 1957).

Anisopteromalus calandrae (Howard)

(= *Aplastomorpha calandrae* (Howard)

Distribution: cosmopolitan.

Interceptions:

Argentina: maize, sorghum
Brazil: maize, cocoa cake
Guyana: rice bran
Mexico: maize
Peru: cottonseed meal, ship's residues
Nigeria: soya beans, oilcake, cocoa beans, beans
West Africa: cocoa beans
Rumania: wheat
Algeria: carobs
Cyprus: carobs
Burma: oilcake, mustard seed
Hong Kong: dried noodles
India: tobacco, dried ginger, turmeric
Pakistan: oilcake

Australia: wheat
Samoa: cocoa beans

Biology: This species is usually found in association with stored products. It has been recorded from a number of stored products pests, including *Sitophilus oryzae*, *Lasioderma serricorne*, *Araecerus fasciculatus*, Bruchidae, *Sitotroga cerealella* and *Ectomyelois ceratoniae*.

Aplastomorpha calandrae (Howard), *see Anisopteromalus calandrae* (Howard)

Bruchobius spp., see *Dinarmus* spp.

Cerocephala dinoderi Gahan

Distribution: United States Department of Agriculture (1974) records it for the first time in Hawaii as well as citing previous finds in the Philippines and Java.

Interception:
Australia: wheat infested with *Sitophilus oryzae*

Biology: this species has been recorded as a parasite of *Dinoderus minutus* in the Philippines and of *Sitophilus oryzae* in Java

Chaetospila elegans Westwood

Distribution: cosmopolitan, in tropical regions and also in warehouses in temperate regions. Loosjes (1957) records that it is able to survive the winter in stored wheat in Holland, at temperatures as low as $-3.5°C$.

Interceptions:
Argentina: wheat, sorghum
Brazil: broken rice
Guyana: rice
Nigeria: cocoa
Burma: rice
Australia: wheat, sorghum

Biology: this is a warehouse species, a parasite of many internal-feeding stored products pests. The Ministry's reference collection contains specimens that have been bred from the following hosts: *Sitophilus oryzae*, *S. granarius*, *Ptinus tectus*, *Stegobium paniceum* and *Alphitobius diaperinus*. De V. Graham (1969) records it from Scolytidae from certain trees in British Honduras and Panama Canal zone. Sharifi (1972) cites records from the following hosts:- *Sitophilus oryzae*, *S. zeamais*, *S. granarius*, *Sitona linearis*, *Caulophilus oryzae*, *Stegobium paniceum*, *Lasioderma serricorne*, *Rhyzopertha dominica*, *Bruchus quadrimaculatus*, *Callosobruchus chinensis*, *C. maculatus* and *Sitotroga cerealella*.

Dinarmus colemani (Crawford)

Distribution: India, South East Asia (British Museum (Natural History)).

Interceptions:
Malaya: moong beans
Singapore: gram

Dinarmus basalis (Rondani)

(= *Dinarmus laticeps* (Ashmead))

Distribution: cosmopolitan.

Interceptions:
  Ethiopia: lentils
  Kenya: lima beans
  Nigeria: beans, cowpeas
  India: gram

Biology: as a parasite of various bruchids, it is known nowadays from many warmer parts of the world (Boŭček, 1974).

Dinarmus laticeps (Ashmead), see *Dinarmus basalis (Rondani)*

Habrocytus cerealella (Ashmead), see *Pteromalus semotus* (Walker)

Habrocytus semotus (Walker), see *Pteromalus semotus* (Walker)

Pteromalus semotus (Walker)

(= *Habrocytus cerealella* Ashmead)

(= *Habrocytus semotus* (Walker))

Distribution: in addition to the countries listed below, the British Museum (Natural History) holds records of this species from the USA, West Indies and Europe.

Interceptions:
  Argentina: wheat, maize
  Australia: wheat
  Brazil: maize in store
  Russia: maize in store

The Ministry's stored products pest reference collection also contains three specimens labelled 'Belfast'

Lariophagus distinguendus (Förster)

Distribution: cosmopolitan.

Interceptions:
  Argentina: wheat, maize, sorghum, rice
  Peru: cottonseed meal
  Uruguay: wheat
  South Africa: maize
  Italy: macaroni
  India: groundnut meal
  Sarawak: illipenuts
  Australia: wheat, sorghum, canary seed, flour

Biology: this is a parasite of beetles associated with stored products, such as *Sitophilus* spp., *Stegobium paniceum*, *Lasioderma serricorne*, Ptinidae etc. Normally only one egg is laid on the host and the parasite larva feeds as an external parasite. Average life cycles of 17 to 18 days are recorded on *S. oryzae* and about 20 days on *Rhyzopertha* and *Stegobium*.

Apocrita — Aculeata

The division of the Apocrita into the two sections, Parasitica and Aculeata, is not very clear cut. The Aculeata contains all of the highly organised and social Hymenoptera. Thus the ants, bees and wasps fall into this group, but so also do the Bethylidae, a family of parastic Hymenoptera which exhibit no social behaviour and which, in their life style, would seem to resemble the members of the Parasitica.

Bethylidae

The larvae of this family live mostly as external parasites on the bodies of Coleoptera or, less often, Lepidoptera larvae. No true nests are built but females have been observed carrying the bodies of their victims to a sheltered place before depositing their eggs on them and in this way they may be said to exhibit a limited degree of maternal care (Perkins, 1976).

Cephalonomia tarsalis (Ashmead)

Distribution: cosmopolitan.

Interceptions:

Argentina: wheat, maize
Bolivia: Brazil nuts
Uruguay: wheat
Nigeria: cocoa beans, groundnuts
Rhodesia: maize
Tanzania: maize meal
Greece: currants
Iran: dried fruit
Iraq: dates
Afghanistan: dried fruit
Burma: rice
India: groundnuts, groundnut meal
Australia: wheat, barley, rye

Biology: this is a warehouse species, a parasite of beetles in stored grain and dried fruit, especially of *Oryzaephilus* spp. (Perkins, 1976). Finlayson (1950) also records it from *Tribolium castaneum* and *Sitophilus oryzae*.

Cephalonomia waterstoni Gahan

Distribution: cosmopolitan.

Interception:

Australian wheat

Biology: this is a warehouse species, a parasite of the larvae of *Cryptolestes* in stored grain (Perkins, 1976). Studies of its biology have been made by Rilett (1949) and Finlayson (1950). The adult female attacks and stings the host larvae, carrying it to a sheltered position before laying one or two, rarely three, eggs on it. The developing larvae feed as external parasites. The life cycle takes about two to three weeks according to temperature.

Holepyris hawaiiensis (Ashmead)

Distribution: cosmopolitan. (Perkins, 1976).

Interceptions:
Nigeria: groundnuts
West Africa: cocoa beans
Solomon Islands: cocoa beans
India: groundnut meal
Indonesia: illipenuts

Biology: a warehouse species, parasitic on the larvae of Phycitinae (Perkins, 1976). Finlayson (1950) records it from *Ephestia elutella* and *Plodia interpunctella*.

Holepyris sylvanidis (Brèthes)

(= *Rhabdepyris zeae* (Turner and Waterson))

Distribution: cosmopolitan (Perkins, 1976).

Interceptions:
East Africa: cottonseed
Kenya: cottonseed cake
Nigeria: groundnuts
Rhodesia: maize
Sudan: cottonseed
Togoland: cottonseed
West Africa: groundnuts
Burma: groundnut cake, maize

Biology: a warehouse species, probably a parasite of *Tribolium* (Perkins, 1976). Finlayson (1950) records it from *Tribolium confusum*.

Plastanoxus monroi Richards

Distribution: West Africa.

Interceptions:
West Africa: cocoa beans
Ghana: cocoa beans
Nigeria: cocoa beans

Biology: according to Perkins (1976) this species is probably parasitic on *Cryptolestes* larvae. The specimens recorded here from Ghanaian and Nigerian cocoa beans, however, were found with an infestation of *Ephestia cautella* larvae, a large number of which were parasitised. Finlayson (1950) records it from *Plodia interpunctella*.

Rhabdepyris zeae (Turner and Waterson), see *Holepyris sylvanidis* (Brèthes)

References

ABDEL-RAHMEN, H. A. (1971) Some factors influencing the abundance of the Indian meal moth, *Plodia interpunctella* Hb. on stored shelled corn (Lepidoptera: Phycitidae). Bull. Soc. ent. d'Egypte **55**: 321–330.

AITKEN, A. D. (1975) Insect travellers. Vol. 1 Coleoptera. Tech. Bull. 31. Ministry of Agriculture, Fisheries and Food, H.M.S.O. xvi + 191pp.

ARBOGAST, R. T. (1976) Suppression of *Oryzaephilus surinamensis* (L.) (Coleoptera, Cucujidae) on shelled corn by the predator *Xylocoris flavipes* (Reuter) (Hemiptera, Anthocoridae). J. Georgia ent. Soc. **11**(1): 67–71.

AVIDOV, Z. and GOTHILF, S. (1960) Observations on the honeydew moth (*Cryptoblabes gnidiella* Milliere) in Israel. 1. Biology, phenology and economic importance. Ktavim (Engl. Ed.) **10** (3–4): 109–124.

AWADALLAH, K. T. and TAWFIK, M. F. S. (1972) The biology of *Xylocoris* (= *Piezostethus*) *flavipes* (Reut.) (Hemiptera: Anthocoridae). Bull. Soc. ent. d'Egypte **56**: 177–189.

BADONNEL, A. (1943 reprinted 1970) Faune de France. Psocoptères. Federation Francaise des Sociétés de Sciences Naturelles.

BALACHOWSKY, A. S. (1972) Entomologie appliquée à l'agriculture. 2. Lepidoptera, Vol. 2. Zygaenoidea, Pyraloidea, Noctuoidea. Masson et Cie, Paris.

BEATSON, S. H. and DRIPPS, J. S. (1972) Long-term survival of cockroaches out of doors. Environmental Health **80** (10): 340–341.

BEHURA, B. K. (1956) The biology of the European earwig, *Forficula auricularia* Linn. Ann. Zool. **1** (5): 117–142.

BELL, C. H. (1975) Effects of temperature and humidity on the development of four pyralid moth pests of stored products. J. stored Prod. Res. **11** (3): 167–175.

BELL, C. H. (1976) Factors governing the induction of diapause in *Ephestia elutella* and *Plodia interpunctella* (Lepidoptera). Physiol. Entom. **1**: 83–91.

BELL, C. H. and WALKER, D. J. (1973) Diapause induction in *Ephestia elutella* (Hübner) and *Plodia interpunctella* (Hübner) (Lepidoptera, Pyralidae) with a dawn-dusk lighting system. J. stored Prod. Res. **9** (3): 149–158.

BELL, C. H. and BOWLEY, C. R. (1980) Effect of photoperiod and temperature on diapause in a Florida strain of the tropical warehouse moth *Ephestia cautella*. J. Insect Physiol. **26**: 533–538.

BELL, C. H., COX, P. D., ALLEN, L. P., PEASON, J. and BEIRNE, M. A. (In Press) Diapause in twenty populations of *Ephestia cautella* from different parts of the world. J. stored Prod. Res.

BENDER, E. (1941) Untersuchungen zur Biologie und Morphologie der in Weinkellern lebenden Kleinschmetterlinge. Z. angew. Ent. **27** (4): 541–584.

BHARADWAJ, R. K. (1966) Observations on the bionomics of *Euborellia annulipes* (Dermaptera: Labiduridae). Ann. ent. Soc. Amer. **59** (3): 441–450.

BOLDT, P. E. (1974) Effects of temperature and humidity on development and oviposition of *Sitotroga cerealella* (Lepidoptera: Gelechiidae). J. Kans. ent. Soc. **47** (1): 30–36.

BOUČEK, Z. (1974) On the Chalcidoidea (Hymenoptera) described by C. Rondani. Estratto da Redia, Vol. LV: 241–285.

BOWDEN, J. and Phipps, J. (1968) Cockroaches (*Periplaneta americana*) as predators. Ent. mon. Mag. **103** (1967) (1238–1240): 175–176.

BRETT, G. A. (1973) The oriental cockroach out of doors. Pest Infestation Control 1968: **70**: 23–25.

BRINDLE, A. (1970) The Dermaptera of the Solomon Islands. Pacific Insects **12** (3): 641–700.

BURGES, H. D. and HASKINS, K. P. F. (1964) Life cycle of the tropical warehouse moth *Cadra cautella* (Wlk.) at controlled temperatures and humidities. Bull. ent. Res. **55** (4): 775–789.

BUSVINE, J. R. (1955) Simple methods of rearing the cricket (*Gryllulus domesticus* L.) with some observations on speed of development at different temperatures. Proc. R. ent. Soc. Lond. (A) **30** (1–3): 15–18.

CALDERON, M., NAVARRO, S. and DONAHAYE, E. (1969) *Ectomyelois ceratoniae* (Zell.) (Lep., Phycitidae) a major pest of stored almonds in Israel. J. stored Prod. Res., **5** (4): 427–428.

CARAYON, J. (1972) Le genre *Xylocoris:* Subdivision et especes nouvelles (Hem. Anthocoridae). Ann. Soc. ent. Fr. (N.S.) **8** (3): 579–606.

CARVALHO, J. C. M. (1956) Insects of Micronesia. Heteroptera: Miridae. Insects of Micronesia **7** (1): 1–100.

CATLING, H. D. (1970) Notes on new minor pests of citrus in South Africa. South African Citrus Journal **444** (11): 13–14.

le CATO, G. L. and DAVIS, R. (1973) Preferences of the predator *Xylocoris flavipes* (Hemiptera: Anthocoridae) for species and instars of stored products insects. Fla. Ent. **56** (1): 57–59.

CHAPMAN, A. J., NOBLE, L. W., ROBERTSON, O. T. and FIFE, L. C. (1960) Survival of the pink bollworm under various cultural and climatic conditions. Prod. Res. Rep. U.S. dep. Agric. **34**: 21pp.

COLE, D. B. and COX, P. D. (1981) Studies on three moth species in a Scottish port silo, with special reference to overwintering *Ephestia kuehniella*. J. stored Prod. Res. **17**: 163–181.

COLYER, C. N. and HAMMOND, C. O. (1968) Flies of the British Isles. Frederick Warne & Co. Ltd., London: 384pp.

COOKE, J. A. L. (1968) A further record of predation by cockroaches (*Periplaneta americana* (L.)). Ent. mon. Mag. **104** (1244–1246): 72.

CORBET, A. S. and TAMS, W. H. T. (1943) Keys for the identification of the Lepidoptera infesting stored food products. Proc. zool. Soc. Lond. (B) **113**: 55–148.

CORBET, S. A. and ROTHERHAM, S. (1965) The life history of the ichneumonid *Nemeritis* (*Devorgilla*) *canescens* (Gravenhorst) as a parasite of the Mediterranean flour moth, *Ephestia* (*Anagasta*) *kuehniella* Zeller under laboratory conditions. Proc. R. ent. Soc. Lond. (A) **40** (4–6): 67–72.

CORNES, M. A. (1973) A check list of the insects associated with stored products in Nigeria. Tech. Rep. 11. Nigerian stored Products Research Institute 1971: 73–98.

COX, P. D. (1974) The influence of temperature and humidity on the life cycles of *Ephestia figulilella* Gregson and *Ephestia calidella* (Guenée) (Lepidoptera: phycitidae). J. stored Prod. Res. **10**: 43–55.

COX, P. D. (1975a) The suitability of dried fruits, almonds and carobs for the development of *Ephestia figulilella* Gregson, *E. calidella* (Guenée) and *E. cautella* (Walker) (Lepidoptera: Phycitidae). J. stored Prod. Res. **11** (3): 229–233.

COX, P. D. (1975b) The influence of photoperiod on the life cycles of *Ephestia calidella* (Guenée) and *Ephestia figulilella* Gregson (Lepidoptera: Phycitidae). J. stored Prod. Res. **11** (2): 75–85.

COX, P. D. (1976) The influence of temperature and humidity on the life cycle of *Ectomyelois ceratoniae* (Zeller) (Lepidoptera: Phycitidae). J. stored Prod. Res. **12** (2): 111–117.

COX, P. D. (1979) The influence of photoperiod on the life-cycle of *Ectomyelois ceratoniae*. J. stored Prod. Res. **15**: 111–115.

COX, P. D., CRAWFORD, L. A., GJESTRUD, G., BELL, C. H. and BOWLEY, C. R. (1981a). The influence of temperature and humidity on the life-cycle of *Corcyra cephalonica*. Bull. ent. Res. **71**: 171–181.

COX, P. D. MFON, M., PARKIN, S. and SEAMAN, J. E. (1981b) Diapause in a Glasgow strain of the flour moth, *Ephestia kuehniella*. Physiol. Ent. **6**: 349–356.

EL-HUSSEINI, M. M. and TAWFIK, M. F. S. (1972) The nutritional effect of animal and plant diets on the development and fecundity of *Euborellia annulipes* (Lucas) (Dermaptera: Labiduridae). Bull. Soc. ent. Egypte **55**: 219–229.

FABER, W. (1978) Die Getreidemotte (*Sitotroga cerealella* Oliv.) jetzt in Österreich auch Freilandschädling. Pflanzenarzt **31** (2): 10.

FALLIS, A. M. (1942) The life cycle of *Apanteles carpatus* (Say) (Hymenoptera: Braconidae), a parasite of the webbing clothes moth, *Tineola biselliella* Hum. Canad. J. Res. (D) **20** (1): 13–19.

FERRIERE, C. and KERRICH, G. H. (1958) Handbooks for the identification of British insects **8** (2a). Hymenoptera, Chalcidoidea. R. ent. Soc. Lond.: 40pp.

FINLAYSON, L. H. (1949) The life history and anatomy of *Lepinotus patruelis* Pearman (Psocoptera, Atropidae). Proc. zool. Soc. Lond. **119**: 301–323.

FINLAYSON, L. H. (1950) The biology of *Cephalonomia waterstoni* Gahan (Hym. Bethylidae), a parasite of *Laemophloeus* (Col. Cucujidae). Bull. ent. Res. **41** (1): 79–97.

FISHER, R. C. (1959) Life history and ecology of *Horogenes chrysostictos* Gmelin (Hymenoptera, Ichneumonidae) a parasite of *Ephestia sericarium* Scott (Lepidoptera, Phycitidae). Canad. J. Zool. **37** (4): 429–446.

FOOD and AGRICULTURE ORGANISATION (1973) Outbreaks and new records. F.A.O. Plant Prot. Bull. **21** (5): 115–118.

FORSYTH, J. (1966) Agricultural insects of Ghana. Ghana Univ. Press, Accra.

FREEMAN, J. A. (1962) The influence of climate on insect populations of flour mills, XII Int. Congr. Ent. Vienna, 1960: 301–308.

FRILLI, F. (1965) Studi sugli imenotteri ichneumonidi. 1. *Devorgilla canescens* (Grav.). Entomologica **1**: 119–209.

GEIGY, R. and ZINKERNAGEL, R. (1941) Beobachtungen beim Aufbau einer technischen Grosszucht der Kleidermotte (*Tineola bisselliella*). Mitt. schweiz. ent. Ges. **18** (4–5): 213–232.

GHOURI, A. S. K. and McFarlane, J. E. (1958) Observations on the development of crickets. Canad. Ent. **90** (3): 158–165.

GOTHILF, S. (1969) The biology of the carob moth *Ectomyelois ceratoniae* (Zell.) in Israel. II. Effect of food, temperature and humidity on development. Israel J. Ent. **4** (1): 107–116.

GOTHILF, S. (1970) The biology of the carob moth *Ectomyelois ceratoniae* (Zell.) in Israel. III Phenology on various hosts. Israel J. Ent. **5:** 161–175.

de V. GRAHAM, M. W. R. (1969) The Pteromalidae of North Western Europe (Hymenoptera: Chalcidoidea). Bull. Br. Mus. nat. Hist. Ent. Suppl. **16:** 908pp.

GREWAL, S. A. and ATWAL, A. S. (1969) The influence of temperature and humidity on the development of *Sitotroga cerealella* Olivier (Gelechiidae, Lepidoptera). J. Res. Punjab Agric. Univ. **6** (2): 353–358.

GRISWOLD, G. H. (1944) Studies on the biology of the webbing clothes moth (*Tineola bisselliella* Hum.). Mem. Cornell agric. Exp. Sta. **262:** 59pp.

GURNEY, A. B. (1953) Distribution, general bionomics and recognition characters of two cockroaches recently established in the United States. Proc. U.S. nat. Hist. Mus. **103** (3315): 39–56.

Guthrie, D. M. and TINDALL, A. R. (1968) The biology of the cockroach. Edward Arnold Ltd., London: 408pp.

HAGSTRUM, D. W. and Sharp, J. E. (1975) Population studies on *Cadra cautella* in a citrus pulp warehouse with particular reference to diapause. J. econ Ent. **68:** 11–14.

HAMMAD, S. M., SHENOUDA, M. G. and EL-DEEB, A. L. (1969) Studies on the biology of *Sitotroga cerealella* Oliv. (Lepidoptera: Gelechiidae). Bull. Soc. ent. d'Egypte **51:** 257–268.

HASSAN, A. A. G., HASSANEIN, M. H. and KAMEL, A. H. (1962) Biological studies on the Indian meal moth *Plodia interpunctella* Hbn. (Lepidoptera, Phycitidae). Bull Soc. ent. d'Egypte **46:** 233–256.

HASSANEIN, M. A. and KAMEL, A. H. (1965) Biological studies on the Mediterranean flour moth *Anagasta kuehniella* (*Ephestia kuehniella*) Zeller (Lepidoptera: Phycitidae). Bull. Soc. ent. d'Egypte **49:** 327–358.

HERRING, J. L. (1967) Insects of Micronesia; Heteroptera: Anthocoridae. Insects of Micronesia 7 (8): 391–414.

HICKIN, N. E. (1974) Household insects pests (Second Edition). Hutchinson and Co. London: 176pp.

HINCKLEY, A. D. (1963) Trophic records of some insects, mites and ticks in Fiji. Bull. Dep. Agric. Fiji, **45.**

HINTON, H. E. (1956) The larvae of the species of Tineidae of economic importance. Bull. ent. Res. **47** (2): 251–346.

HÖLLER, G. (1965) Über die Wollverdauung der Raupen von *Hofmannophila pseudospretella* (Lepidopt. Oecophoridae). Anz. Schädlingsk. **38** (3): 36–39.

HOWE, R. W. (1952) Entomological problems of food storage in Northern Nigeria. Bull. ent. Res. **43** (1): 111–144.

HOWE, R. W. and FREEMAN, J. A. (1955) Insect infestation of West African produce imported into Britain. Bull. ent. Res. **46** (3): 643–668.

HURLOCK, E. T. (1959) Recent occurrences of *Mussidia nigrivenella* Rag. (Lep. Phycitidae) on products imported into Britain. Ent. mon. Mag. **95** (1141): 128.

HURLOCK, E. T. (1963) The infestation of Canadian produce inspected in United Kingdom ports between 1953 and 1959. Can. Ent. **95** (12): 1263–1284.

HURLOCK, E. T. (1964) Infestation of foodstuffs from the United States of America inspected in the United Kingdom between 1953 and 1961. Bull. ent. Res. **55** (1): 173–192.

HUSSAIN, S. and ASLAM, N. A. (1970) Some observations on a beneficial Reduviid bug: *Amphibolus venator* Klug. (Fam. Reduviidae: Hemiptera). Agriculture Pakistan **21** (1): 37–42.

IMMS, A. D. (1957) A general textbook of entomology. 9th Edition. Methuen, London.

JACOB, T. A. and COX, P. D. (1977) The influence of temperature and humidity on the life cycle of *Ephestia kuehniella* Zeller (Lep. Pyralidae). J. stored Prod. Res. **13:** 107–118.

JAY, E., DAVIS, R. and BROWN, S. (1968) Studies on the predacious habits of *Xylocoris favipes* (Reuter) (Hemiptera: Anthocoridae). J. Ga. ent. Soc. **3** (3): 126–130.

JEANNEL, R. (1960) Introduction to entomology (Translated by H. Oldroyd). London. Hutchinson, 344pp.

KAMEL, A. H. and HASSANEIN, M. H. (1967) Biological studies on *Corcyra cephalonica* Staint. (Lepidoptera: Galleriidae). Bull. Soc. ent. d'Egypte **51:** 175–196.

KHAN, N. H. (1954) Ecological observations on the eggs of *Gryllodes sigillatus* (Walker). Indian J. Ent. **16** (1): 24–26.

KLOSTERMEYER, E. C. (1942) The life history and habits of the ring-legged earwig *Euborellia annulipes* (Lucas) (Order Dermaptera). J. Kans. ent. Soc. **15** (1): 13–18.

KURSTAK, E. (1964) Niektóre dane z biologii i ekologii peczola mklikowca (*Nemeritis canescens* Grav., Ichneumonidae, Hymenoptera). Pr. nauk. Inst. Ochr. Rosl. **6** (1): 173–187.

References

LEGNER, E. F. and EASTWOOD, R. E. (1973) Seasonal and spatial distribution of *Tinea fuscipunctella* on poultry ranches. J. econ. Ent. **66** (3): 685–687.

LOOSJES, F. E. (1957) Ervaringen met *Chaetospila elegans* (Westw.) (Hymenoptera; Pteromalidae) een parasiet van enige soorten voorraadinsecten. Ent. Ber. **17** (4): 74–76.

LUKEFAHR, M. J., NOBLE, L. W. and MARTIN, D. F. (1964) Factors influencing diapause in the pink bollworm. Techn. Bull. U.S. Dept. Agric. **1304**: 17pp.

McFARLANE, J. A. (1963) An annotated record of Coleoptera, Lepidoptera, Hemiptera and Hymenoptera associated with stored produce in Jamaica. Trop. Agric. **40** (3): 211–216.

MATTHEWS, R. W. (1974) Biology of Braconidae. Ann. Rev. Ent. **19**: 15–32.

MELAMED-MADJAR, V. (1971) Bionomics and ecology of the earwig *Anisolabis (Euborellia) annulipes* (Luc.) (Labiduridae — Dermaptera) in Israel. Z. Angew. Ent. **69** (2): 170–176.

MICHAEL, P. (1968) The carob moth, J. Dep. Agric. West. Aust. **9** (2): 81–82.

MORERE, J. L. and LE BERRE, J. R. (1967) Etude au laboratoire du developpement de la pyrale *Plodia interpunctella* (Hübner) (Lep. Phycitidae). Bull. Soc. ent. Fr. **72** (5–6): 157–166.

MORRILL, A. W. (1942) Notes on the biology of *Microbracon hebetor*. J. econ. Ent. **35** (4): 593–594.

NEISWANDES, C. R. (1944) The ring-legged earwig *Euborellia annulipes* (Lucas). Bull. Ohio agric. Exp. Sta. **648**: 14pp.

NOTINI, G. (1939) Klädesmalen. Medd. St. Växtskyddsanst. **28**: 32pp.

OLDROYD, H. (1964) The natural history of flies. Weidenfield & Nicolson, London.

OLDROYD, H. (1977) The suborders of Diptera. Proc. ent. Soc. Wash. **79** (1): 3–10.

OMAR, M. T. A., KAMEL, A. H., EL-KIFL, A. H. and WAHAB, A. E. A. (1974) Ecology and biological studies on *Cadra calidella* (Guen.) a pest attacking dry dates in Egypt (Lepidoptera: Phycitidae). Bull. Soc. ent. d'Egypte **57**: 361–370.

OSTROVSKII, N. L. (1940) A bio-chemical method for the determination of the death of the larvae of *Systole ceriandri* Nik. after fumigation of seeds. Bull. Plant. Prot. **4**: 53–56.

Le PELLEY, R. H. (1959) Agricultural insects of East Africa. Nairobi: East Africa High Commission, x + 307pp.

PERKINS, J. F. (1976) Handbooks for the identification of British insects. **6** (3a): Hymenoptera, Bethyloidea. R. ent. Soc. Lond.: 38pp.

PIMENTEL, D. and CRANSTON, F. (1960) The house cricket, *Acheta domestica* and the house fly, *Musca domestica* as a model predator-prey system. J. econ. Ent. **53** (1): 171–172.

PINGALE, S. V. (1954) Biological control of some stored grain pests by the use of a bug predator, *Amphibolus venator* Klug. Indian J. Ent. **16** (3): 300–302.

POPE, P. (1953) Studies of the life histories of some Queensland Blattidae (Orthoptera). I. The domestic species. Proc. roy. Soc. Qd. **63**: 23–46.

PRESS, J. W., FLAHERTY, B. R. and ARBOGAST, R. T. (1974) Interactions among *Plodia interpunctella*, *Bracon hebetor* and *Xylocoris flavipes*. Environ. Ent. **3** (1): 183–184.

PREVETT, P. F. (1968) Some laboratory observations on the life cycle of *Cadra calidella* (Guen.) (Lepidoptera: Phycitidae). J. stored Prod. Res. **4** (3): 233–238.

PUNJ, G. K. (1967) Dietary efficiency of natural foods for the growth and development of *Tribolium castaneum* Herbst. and *Corcyra cephalonica*. Bull. Grain Technol. **5** (4): 209–213.

RAGGE, D. R. (1965) Grasshoppers, crickets and cockroaches of the British Isles. Wayside and Woodland Series, Frederick Warne, London: 299pp.

RAO, D. S. (1954) Notes on the rice moth, *Corcyra cephalonica* (Stainton), (Family: Galeriidae) Lepidoptera. Indian J. Ent. **16** (2): 95–114.

RICCI, M. (1964) Su alcuni aspetti della biologia di *Blatta orientalis* L. Riv. parassit. **24** (1963)(3): 185–198.

RICHARDS, O. W. and WALOFF, N. (1946) The study of a population of *Ephestia elutella* Hübner (Lep. Phycitidae) living on bulk grain. Trans. R. ent. Soc. Lond. **97** (11): 253–298.

RILEET, R. O. (1949) The biology of *Cephalonomia waterstoni* Gahan. Canad. J. Res. (D) **27** (3): 93–111.

RIVNAY, E. and ZIV, M. (1963) A contribution to the biology of *Gryllus bimaculatus* Deg. in Israel. Bull. ent. Res. **54**: 37–43.

ROBINSON, G. S. (1979) Clothes-moths of the *Tinea pellionella* complex: a revision of the world's species (Lepidoptera: Tineidae) Bull. Br. Mus. nat. Hist. (Ent.) **38** (3): 57–128.

ROTH, L. M. and WILLIS, E. R. (1960) The biotic associations of cockroaches. Smithson. misc. Collns. **141**: 470pp.

RUPPEL, R. F. (1977) An inventory of stored grain insects in Michigan. Great Lakes Entomol. **10**: 243–244.

RYCKMAN, R. E. and RYCKMAN, A. E. (1967) Epizootiology of *Trypanosoma cruzi* in southwestern North America. Part XI, Biology of the genus *Reduvius* in North America and the possible relationship of *Reduvius* to the epizootiology of *Trypanosoma cruzi* (Hemiptera: Reduviidae) (Kinetoplastida: Trypanosomidae). J. med. Ent. **4** (3): 326–334.

SALT, G. (1976) The hosts of *Nemeritis canescens* a problem in the host specificity of insect parasitoids. Ecol. Ent. **1** (1): 63–67.

SAVOV, D. (1973) Development of *Plodia interpunctella* Hb. (Lepidoptera, Pyralidae) in the optimum temperature range. Gradinarska i Lozarska Nauka **10** (5): 33–40.

SEYMOUR, P. R. (1979) Invertebrates of economic importance. Common and scientific names. Ministry of Agriculture, Fisheries and Food. H.M.S.O. viii + 132pp.

SHAHJAHAN, M. (1974) Extent of damage of unhusked store rice by *Sitotroga cerealella* Oliv. (Lepidoptera: Gelechiidae) in Bangladesh. J. stored Prod. Res. **10** (1): 23–26.

SHARIFI, S. (1972) Radiographic studies of the parasite *Chaetospila elegans* on the maize weevil *Sitophilus zeamais*. Ann. ent. Soc. Amer. **65** (4): 825–856.

SHIRES, S. W. (1976) Observations on *Liposcelis bostrychophilus* Badonnel (Insecta, Psocoptera) a pest of laboratory insect cultures. Tropical Stored Products Centre, Private Communication.

SIDDIQUI, W. H. and BARLOW, C. E. (1973) Population growth of *Anagasta kuehniella* (Lepidoptera: Pyralidae) at constant and alternating temperatures. Ann. ent. Soc. Amer. **66** (3): 579–585.

SKUHRAVY, V. (1960) Die Nahrung des Ohrwurms (*Forficula auricularia* L.) in den Feldkulturen. Acta Soc. ent. Csl. **57** (4): 329–339.

SMITH, K. G. (1956) The occurrence and distribution of *Aphomia gularis* (Zell.) (Lep. Galleriidae) a pest of stored products. Bull. ent. Res. **47** (4): 655–667.

SMITH, K. G. (1960) Insect infestation associated with French shelled walnuts with particular reference to the occurrence of *Aphomia gularis* (Zell.) (Lep., Galleridae). Bull. ent. Res. **50** (4): 711–716.

SMITH, K. G. (1965) Some aspects of the biology of *Paralipsa (Aphomia) gularis* (Zell.) in relation to its distribution. Proc. XII Int. Congr. Ent. Lond. **8** (16): 626.

SOLIMAN, H. S. (1942) Studies in the biology of *Microbracon hebetor* Say (*Hymenoptera: Braconidae*). Bull. Soc. Fouad Ent. **24:** 215–247.

SOLOMON, M. E. and ADAMSON, B. E. (1955) The powers of survival of storage and domestic pests under winter conditions in Britain. Bull. ent. Res. **46:** 311–355.

SOUTHGATE, B. J. (1971) On the identity of *Caryedon pallidus* (Olivier) (Col. Bruchidae), and the description of two new *Caryedon* Spp. Bull. ent. Res. **60:** 409–414.

SOUTHWOOD, T. R. E. and LESTON, D. (1959) Land and water bugs of the British Isles. Wayside and Woodland Series, Frederick Warne, London.

STRÜMPEL, H. (1964) Physiologische und histologische Untersuchungen zur Diapause bei *Ephestia elutella* Hübner (Lep. Phycitidae). Mitt. Hamburg, zool. Mus. **61:** 199–245.

ŠTYS, P. (1973) Cases of facultative parasitism of Lyctocorinae (Heteroptera: Anthocoridae) on man in Czechoslovakia. Folia Parasitologica **20** (1): 103–104.

SURLIKAR, V. V. and TEMBE, V. B. (1969) Bionomics of *Amphibolus venator* (Klug) a reduviid predator on insect pests of stored products. J. Bombay nat. Hist. Soc. **66** (3): 640–645.

TAHER, EL-SAYED, M. and ABDEL-RAHMAN, H. A. (1960) On the biology and life history of the pink bollworm *Pectinophora gossypiella* (Saunders) (Lepidoptera: Gelechiidae). Bull. Soc. ent. d'Egypte **44:** 71–90.

TAWFIK, M. F. S. and EL-HUSSEINI, M. M. (1971) The life history of *Xylocoris* (=*Piezostethus*) *galactinus* (Fieber) (Hemiptera: Anthocoridae). Bull. Soc. ent. d'Egypte **55:** 171–183.

TEMBE, V. B. and SURLIKAR, V. V. (1968) *Amphibolus venator* (Klug) (Hemiptera, Heteroptera) a predator on insect pests of stored products. Bull. Grain Tech. **6** (1): 24–26.

TSUJI, H. (1963) Experimental studies on the larval diapause of the Indian meal moth, *Plodia interpunctella* Hübner (Lep. Pyralidae). Thesis Kyusha University: 88pp.

TZANAKAKIS, M. E. (1959) An ecological study of the Indian meal moth *Plodia interpunctella* (Hübner) with emphasis on diapause. Hilgardia **29** (5): 205–246.

U.S.D.A. (1974) Hawaiian insect report. Cooperative Insect Report **24** (23): 417.

VUKASOVIĆ, P. (1940) Contribution à l'étude de *Sitotroga cerealella* 01. Arh. Min. Pop'oprivr. **7** (18): 3–49.

WALOFF, N. (1948) Development of *Ephestia elutella* Hb. (Lep. Phycitidae) on some natural foods. Bull. ent. Res. **39** (1): 117–130.

WALOFF, N. (1949) Observations on larvae of *Ephestia elutella* Hübner (Lep., Phycitidae) during diapause. Trans R. ent. Soc. Lond. **100** (5): 147–159.

WHEELER, M. R. and TAKADA, H. (1964) Insects of Micronesia. Diptera: Drosophilidae. Insects of Micronesia **14** (6): 163–242.

WILLIAMS, G. C. (1964) The life history of the Indian meal moth, *Plodia interpunctella* (Hübner) (Lep. Phycitidae) in a warehouse in Britain and on different foods. Ann. appl. Biol. **53** (3): 459–475.

WOODROFFE, G. E. (1951a) A life history study of *Endrosis lactella* (Schiff.) (Lep. Oecophoridae). Bull. ent. Res. **41** (4): 749–760.

WOODROFFE, G. E. (1951b) A life history study of the brown house moth *Hofmannophila pseudospretella* (Staint.) (Lep. Oecophoridae). Bull. ent. Res. **41** (3): 529–553.

WOODROFFE, G. E. (1953) An ecological study of the insects and mites in the nests of certain birds in Britain. Bull. ent. Res. **44** (4): 739–772.

WOODROFFE, G. E. and HALSTEAD, D. G. H. (1959) *Fulvius brevicornis* Reut. (Hem. Miridae) and other insects breeding in stored Brazil nuts in Britain. Ent. mon. Mag. **45**: 130–133.

WOODROFFE, G. E. and SOUTHGATE, B. J. (1951) A common host and habitat of *Apanteles carpatus* Say. (Hym. Braconidae) in Britain. Ent. mon. Mag. **87**: 171.

WYGODZINSKY, P. and USINGER, R. L. (1960) Insects of Micronesia. Heteroptera: Reduviidae. Insects of Micronesia **7** (5): 231–283.

ZAGULYAEV, A. K. (1966) Natural enemies of the lac insect among the Lepidoptera. (In Russian). Zool. Zh. **45** (7): 1033–1039.

Appendix

Insects that were recorded in association with stored products but which were not necessarily pests, predators, or parasites. They were frequently stray introductions.

Orthoptera- Saltatoria

Acrididae

Anacridium aegyptium (L.)
  Distribution: Mediterranean areas of Europe and North Africa.
  Interception: Italy: cabbages, fennel, cauliflowers
Gesonula punctifrons (Stål)
  Distribution: India, China, Sri Lanka, Thailand.
  Interception: tea of unknown origin.
Schistocerca gregaria (Forskål) – – desert locust
  Distribution: North, West, and East Africa, Portugal, Spain, Turkey, Iran, Pakistan and northern India.
  Interception: ship's hold origin unknown.

Gryllidae

Gryllulus commodus Walker, see *Teleogryllus commodus* (Walker)
Homoeogryllus reticulatus (Fabricius)
  Distribution: tropical Africa.
  Interception: Africa: timber origin unknown
Scapsipedus marginatus (Afzelius and Brannius)
  Distribution: Africa.
  Interception: Nigeria: bones
Teleogryllus commodus (Walker) (= *Gryllulus commodus* Walker)
  Distribution: Australia, N. Zealand, N. Caledonia.
  Interception: Australia: twice, tins of peaches.

Gryllotalpidae

Gryllotalpa gryllotalpa L. – – mole cricket
  Distribution: Europe, North Africa, western Asia, rare in Britain.
  Interception: Egypt: kibbled onions,
              Turkey: tomato paste

Tettigonidae

Homorocoryphus nitidulus vicinus (Walker), *see Ruspolia differens* (Serville)
Ruspolia differens (Serville)
(= *Homorocoryphus nitidulus vicinus* (Walker))

  Distribution: Africa.
  Interception: Uganda: soluble tea powder
Nesonotus tricornis (Thunberg)
  Distribution: W. Indies.
  Interception: bananas of unknown origin.

Dermaptera

Cranopygia burri Hincks
  Distribution: southern India.
  Interception: Sri Lanka: desiccated coconut
Euborellia stali (Dohrn)
  Distribution: Indo-Australia and Madagascar.
  Interception: East Africa: coconuts and coconut shell

Dictyoptera

Blattodea

Blatta lateralis (Walker)
(= *Shelfordella tartara* (Saussure))
  Distribution: N. E. Africa, E. Mediterranean, Soviet Central Asia.
  Interception: Iran: sultanas and almonds
                Burma: groundnut cake
Burchellia vinula (Stål,)
(= *Theganopteryx vinula* Stål)
  Distribution: Central and South Africa.
  Interception: South Africa: groundnuts
Dorylaea rhombifolia (Stollmeyer), see *Neostylopyga rhombifolia* (Stollmeyer)
Eurycotis decipiens (Kirby)
(= *Pelmatosilpha decipiens* (Kirby))
  Distribution: West Indies, Central and South America.
  Interception: West Indies: hold of ship after banana cargo
Henschoutedenia epilamproides (Shelford)
  Distribution: West Africa.
  Interception: Cameroons: bananas
Henschoutedenia flexivitta (Walker)
(= *Nauphoeta brazzae* (Bolivar))
  Distribution: Africa.
    Interceptions: on bananas of unknown origin, in store
                on a sack of potatoes in a canteen

Henschoutedenia tectidoma Gurney
  Distribution: West Africa.
  Interception: bananas in store
Holocompsa nitidula (Fabricius)
  Distribution: Central America, West Africa.
  Interception: Brazil: Brazil nuts, assorted produce
    Nigeria: bones
Leucophaea maderae (Fabricius)
  Distribution: West Africa, Madeira, the West Indies, Central and South America.
  Interceptions: South Africa: maize meal, in the hold of a ship from East Africa
Loboptera decipiens (Germar)
  Distribution: southern Europe, eastern Mediterranean, North Africa.
  Interception: Italy: rice
Nauclides nigra (Brunner)
  Distribution: Central America, West Indies.
  Interceptions: bananas and potatoes of unknown origin, in a factory that processes breakfast food from American maize
Nauphoeta brazzae (Bolivar), see *Henschoutedenia flexivitta* (Walker)
Neostylopyga rhombifolia (Stollmeyer)
(= *Dorylaea rhombifolia* (Stollmeyer))
  Distribution: Indo-Malayan and New World tropics, Philippines, Hawaii.
  Interceptions: India: leather
    Japan: crated motorcycle
Nyctibora laevigata (Beauvois)
  Distribution: West Indies.
  Interceptions: South America: bananas
    Jamaica: bananas
Nyctibora noctivaga Rehn
  Distribution: Central America, West Indies.
  Interception: West Indies: bananas
Oxyhaloa ferreti (Reiche and Fairmaire)
  Distribution: Africa.
  Interception: East Africa: chillies
Panchlora nivea (L.)
  Distribution: Central and South America and the West Indies.
  Interception: West Indies: bananas
Pelmatosilpha decipiens (Kirby), see *Eurycotis decipiens* (Kirby)
Pelmatosilpha larifuga Gurney
  Distribution: Central and South America, West Indies.
  Interception: West Indies: bananas
Pelmatosilpha marginalis Brunner
  Distribution: West Indies.
  Interception: on Brazilian bananas in store
Pelmatosilpha micra Hebard

  Distribution: W. Indies, South America.
  Interception: bananas, unknown origin
Periplaneta brunnea (Burmeister) – – brown cockroach
  Distribution: Old and New World tropics.
  Interception: Thailand: bones
Pycnoscelus surinamensis (L.) – – Surinam cockroach
  Distribution: throughout warmer parts of the world.
  Interception: Australia: dried fruit, millet
Shelfordella tartare (Saussure), see *Blatta lateralis* (Walker)
Supella longipalpa (Fabricius) – – brownbanded cockroach
(= *Supella supellectilium* (Serville))
  Distribution: Africa N of equator, Madagascar, S. Asia, Arabia, India, Malaya, New Guinea, Australia, South America, Central America, West Indies, North America, Europe in heated buildings.
  Interception: Cyprus: raisins
Supella supellectilium (Serville), see *Supella longipalpa* (Fabricius)
Theganopteryx vinula Stål, see *Burchellia vinula* (Stål)

Mantodea

Mantis religiosa L. – – praying mantis
  Distribution: S. Europe, Asia, N. Africa.
  Interception: one specimen in aircraft cargo hold

Isoptera

Coptotermes formosanus (Shiraki)
  Distribution: Far East, South Africa, Pacific islands, Hawaii, Sri Lanka, United States.
  Interceptions: South Africa: tinned fruit
    Hong Kong: miscellaneous cargo
Coptotermes heimi (Wassmann)
  Distribution: India, Pakistan.
  Interception: India: cloth
Kalotermes flavicollis (Fabricius)
  Distribution: Distribution: southern Europe, Mediterranean region, North Africa, Asia minor.
  Interception: Greece: currants
Porotermes adamsoni (Froggatt)
  Distribution: coastal districts of south-eastern Australia and in Tasmania.
  Interception: southern Australia: flour

Hemiptera — Homoptera

Cercopidae

Cercopis sanguinea (Geoffroy), see *Cercopis vulnerata* Rossi

Cercopis vulnerata Rossi – – red and black froghopper

(= *Cercopis sanguinea* (Geoffroy))

Distribution: Europe, S. Russia.

Interception: India, ricebran (almost certainly stray from British dockside fauna

Cicadellidae

Nephotettix bipunctatus (Fabricius), see *Nephotettix virescens* (Distant)

Nephotettix impicticeps Ishikara, see *Nephotettix virescens* (Distant)

Nephotettix modulatus Melichar

Distribution: Africa, madagascar.

Interception: ship's hold

Nephotettix virescens (Distant)

(= *Nephotettix bipunctatus* (Fabricius))

(= *Nephotettix impicticeps* Ishikara)

Distribution: Malaysia, Japan, Philippines, Taiwan, Sri Lanka, India sub continent.

Interception: Burma: cooked rice

Delphacidae

Perkinsiella saccharicida Kirkaldy

Distribution: S. Africa, Madagascar, Mauritius, Reunion, Columbia, Hawaii, Singapore, Queensland.

Interception: Australia: rice

Hemiptera — Heteroptera

Aneuridae see Aradidae

Anthocoridae

Amphiareus constrictus (Stål)

(= *Lasiochilus sladeni* Distant)

Distribution: Madeira, Morocco, Cape Verde Is., tropical Africa, Mascarenes, Madagascar, Antilles, Bahamas, Guyana, Brazil, Micronesia, Indonesia, New Guinea.

Interception: Ghana: cocoa beans, palm kernels

Amphiareus melanicus (Distant)

(= *Buchananiella melanicus* (Distant))

Distribution: Calcutta.

Interception: empty hold after cargo of rice from Madras

Buchananiella melanicus (Distant), see *Amphiareus melanicus* (Distant)

Lasiochilus alluaudi Reuter

Distribution: Seychelles.

Interception: Ghana: palm kernels

Lasiochilus sladeni Distant, see *Amphiareus constrictus* (Stål)

Physopleurella mundula Buchan White

Distribution: Hawaii, Sandwich Is., Seychelles, India.

Interception: Ghana: palm matting around cocoa beans

Aradidae (= Aneuridae)

Mezira proximus (Walker), see *Neuroctemus proximus* (Walker)

Neuroctemus proximus (Walker)

(= *Mezira proximus* (Walker))

Distribution: Australia.

Interception: Australia: timber

Belostomatidae

Hydrocyrius columbiae Spinola

Distribution: Africa.

Interception: West Africa: palm kernels

Lethocerus indicus (Le Peletier and Serville)

Distribution: Oriental and Malayan regions.

Interception: empty hold after cargo from Burma

Cimicidae

Cimex hemipterus (Fabricius)

(= *Cimex rotundatus* (Signoret))

Distribution: Old and New World tropics.

Interceptions: Zanzibar: cloves
India: miscellaneous foodstuffs, ship's cabin

Cimex lectularius L. – – bed bug

Distribution: cosmopolitan.

Interception: Kenya: moong

Cimex rotundatus, see *Cimex hemipterus* (Fabricius)

Cydnidae

Aethus indicus (Westwood)

(= *Cydnus indicus* Westwood)

Distribution: Asia, Australia, Africa.

Interception: East Africa: spillage
Sierra Leone: cocoa beans

Cydnus indicus Westwood, see *Aethus indicus* (Westwood))

Geotomus pygmaeus (Dallas)

Distribution: Sumatra, Borneo, Java, New Guinea, New Caledonia and adjacent islands, Bonin, Mariana, Caroline, Marshall and Gilbert Islands.

Interception: Malaya: in a tin of pineapple

Pangaeus bilineatus (Say)

   Distribution: United States of America, Mexico, Guatemala, Bermuda.

   Interception: United States (Texas): dried salted sheepskins

Pangaeus piceatus Stål

   Distribution: southern Mexico, Guatemala, Costa Rica, Puerto Rica, Colombia, Brazil, Peru.

   Interception: Argentina: sunflower seed meal

Gerridae

Limnogonus nitidus (Mayr)

   Distribution: India, Sri Lanka, Thailand.

   Interception: Burma: groundnut cake

Lygaeidae

Aphanus pini (L.), see *Rhyparochromus pini* (L.)

Aphanus sordidus (Fabricius), see *Elasmolomus sordidus* (Fabricius)

Blissus diplopterus Distant, see *Macchiademus diplopterus* (Distant)

Elasmolomus sordidus (Fabricius)

(= *Aphanus sordidus* (Fabricius))

   Distribution: Philippines, India, Thailand, Burma, south China, Okinawa, Mariana Is.

   Interception: Gambia: groundnuts

Heterogaster urticae (Fabricius) – – nettle ground bug

   Distribution: Palaearctic.

   Interception: Canadian flour, stray

Ischnodemus diplopterus (Distant), see *Macchiademus diplopterus* (Distant)

Macchiademus diplopterus (Distant)

(= *Blissus diplopterus* Distant)

(= *Ischnodemus diplopterus* (Distant))

   Distribution: South Africa.

   Interception: South Africa: nectarines

Rhyparochromus pini (L.)

(= *Aphanus pini* (L.))

   Distribution: Palaearctic.

   Interception: empty ship's hold

Miridae

Fulvius subnitens Poppius

   Distribution: New Guinea.

   Interception: Malaya: sandalwood

Pentatomidae

Aelia acuminata (L.) – – bishop's mitre bug

   Distribution: Europe.

   Interception: empty ship's hold

Dolycoris baccarum (L.) – – sloe bug

   Distribution: Europe.

   Interception: Hungary: wheat

Elasmucha ferrugata (Fabricius)

   Distribution: N. Europe, USSR, Japan.

   Interception: Poland: bilberry flavoured yoghurt

Euschistus impictiventris Stål

(= *Euschistus servus* (Say))

   Distribution: N. America.

   Interception: USA wheat: empty ship's hold

Euschistus servus (Say), see *Euschistus impictiventris* Stål

Nezara viridula (L.) – – green vegetable bug

   Distribution: throughout warm regions of the world.

   Interception: Australia: millet seed

Pyrrhocoridae

Dysdercus cingulatus (Fabricius)

   Distribution: Oriental and Australasian regions.

   Interception: India: cottonseed cake

Pyrrhocoris apterus (L.)

   Distribution: Europe, northern and central Asia.

   Interception: Turkey: bones

Scantius aegyptius (L.)

   Distribution: Mediterranean region.

   Interception: Spain: canned fruit

Scutelleridae

Eurygaster austriaca (Schrank)

   Distribution: Mediterranean region.

   Interception: Hungary: wheat (twice)

Eurygaster integriceps Puton

   Distribution: Italy, Europe, W. Pakistan, southern USSR.

   Interception: Russia: wheat

Eurygaster maura (L.)

   Distribution: central and southern Europe, N. Africa.

   Interceptions: Bulgaria: wheat, Hungary: wheat

Scutiphora pedicellata Kirby

   Distribution: Australia.

   Interception: Australia: hops

Lepidoptera

Blastobasidae

Blastobasis inana (Butler)

Distribution: Hawaiian Is.

Interceptions: Borneo: illipenuts
Sarawak: illipenuts

Blastobasis industria Meyrick
Distribution: East Africa, Ethiopia, Morocco.
Interception: China (via Rotterdam): gallnuts

Cosmopterygidae

Anatrachyntis falcatella (Stainton)
(= *Pyroderces falcatella* (Stainton))
Distribution: India, Sri Lanka, Malaya, Solomon Is, Fiji.
Interception: Indonesia: illipenuts
Pyroderces falcatella (Stainton), see *Anatrachyntis falcatella* (Stainton)

Gelechiidae

Gnorimoschema operculella (Zeller), see *Phthorimaea operculella* (Zeller)
Mometa zemiodes Durrant
(= *Platyedra zemiodes* (Durrant))
(= *Pectinophora zemiodes* (Durrant))
Distribution: Nigeria, Angola, Uganda.
Interception: Eritrea: cottonseed
Pectinophora zemiodes (Durrant), see *Mometa zemiodes* Durrant
Phthorimaea operculella (Zeller) – – potato moth
(= *Gnorimoschema operculella* (Zeller))
Distribution: subtropical and tropical.
Interceptions: Cyprus: potatoes
Israel: potatoes
Italy: sweet almonds, inside a can of tomatoes
Turkey: dried sultanas, dried apricots
Platyedra zemiodes (Durrant), see *Mometa zemiodes* Durrant

Geometridae

Hydrelia risata (Guenée)
Distribution: Australia.
Interception: Australia: sheepskins
Zermizinga indocilisaria Walker
Distribution: Australia: New Zealand.
Interception: Australia: broken rice

Noctuidae

Achaea catocaloides Guenée
Distribution: Africa.
Interception: West Africa: palm kernels
Achaea lienardi (Boisduval)
Distribution: Africa.

Interception: Ghana: cocoa beans
Acontia nivipicta Butler
Distribution: Australia, New Guinea.
Interception: Australia: millet
Apamea sordens (Hufnagel) – – rustic shoulder knot moth
Distribution: north, east & central Europe, Japan, North America east of the Rocky Mountains.
Interception: Denmark: wheat
Caradrina clavipalpis (Scopoli) – – pale mottled willow moth
Distribution: Europe, Near East and Central Asia.
Interception: USA: wheat and maize
Elaphria chalcedonia (Hübner)
Distribution: North America.
Interception: USA: rice
Helicoverpa armigera (Hübner), see *Heliothis armigera* (Hübner)
Heliothis armigera (Hübner) – – Old world bollworm
(= *Helicoverpa armigera* (Hübner))
Distribution: Mediterranean Europe, Africa, tropical Asia, Pacific Is.
Interception: Italy: inside a can of tomatoes
Plusia argentifera Guenée
Distribution: Australia.
Interception: Australia: canary seed

Pyralidae

Chrysauginae

Caphys bilineata Stollmeyer
Distribution: South America, West Indies.
Interception: South America: Brazil nuts

Crambinae

Agriphila straminella Denis and Schiffermüller
(= *Crambus culmella* L.)
Distribution: temperate, Palaearctic regions.
Interception: empty ship's hold
Crambus culmella L., see *Agriphila straminella* Denis and Schiffermüller

Galleriinae

Achroia grisella (Fabricius)
Distribution: cosmopolitan in beehives.
Interceptions: Nigeria: beeswax
Hong Kong: used sacks
Doloessa viridis Zeller
Distribution: Indo-Australian region.

Interceptions: Borneo: illipenuts
Indonesia: illipenuts
Malaya: palm kernels
Singapore: nutmegs

Phycitinae

Cryptoblabes gnidiella Milliere
Distribution: Mediterranean, Africa, India, New Zealand, Madeira, Canary Is., Sumatra, Malaysia.
Interception: Spain: almonds
Ephestia parasitella Staudinger
Distribution: Europe.
Interception: USA: groundnuts
Euzophera semifuneralis Walker
Distribution: North America.
Interception: Canada: under bark of a log

Pyralinae

Aglossa caprealis (Hübner) – – small stable tabby moth
Distribution: cosmopolitan.
Interceptions: Nigeria: cocoa beans
in residues in hold of a ship returning from S. Africa
Aglossa ocellalis Lederer
Distribution: Africa.
Interceptions: Sierra Leone: hooves and horns
West Africa: ginger
Pyralis pictalis (Curtis)
Distribution: Indo-Australia and tropical Africa.
Interception: Sarawak: illipenuts

Pyraustinae

Marasmia poeyalis Boisduval
Distribution: tropical.
Interception: Australia: rice
Ostrinia nubilalis (Hübner) – – European corn borer
(= *Pyrausta nubilalis* (Hüber)
Distribution: mainly northern hemisphere.
Interceptions: Canada: in a tin of sweet corn
Italy: in a tin of peeled tomatoes
Pyrausta nubilalis (Hübner) see *Ostrinia nubilalis* (Hübner)

Pieridae

Colias eurytheme Biosduval
Distribution: southern Canada, USA, and the uplands of Mexico.
Interception: USA: prunes

Tineidae

Amydria vastella (Zeller), see *Ceratophaga vastella* (Zeller)
Ceratophaga vastella (Zeller)
(= *Tinea vastella* Zeller)
(= *Amydria vastella* (Zeller))
Distribution: Africa.
Interceptions: Nigeria: horns, hooves
East Africa: groundnut cake
West Africa: cow bones

Haplotinea ditella (Pierce and Metcalfe)
Distribution: Britain, Holland, Austria, Germany, European Russia, Trans-Caucasia, Central Asia.
Interceptions: Canda: soyabean meal,
wheat and barley residues in the empty hold of a ship on the North Atlantic route,
in residues of French wheat and maize

Lindera tessellatella Blanchard, see *Setomorpha tessellatella* (Blanchard)
Monopis crocicapitella (Clemens)
Distribution: cosmopolitan.
Interceptions: Holland: wheatbran,
in ship's hold with wheat and maize residues.

Nemapogen cloacella (Haworth) – – cork moth
(= *Tinea cloacella* Haworth)
Distribution: cosmopolitan.
Interception: Turkey: walnuts

Opogona astragaloides Meyrick
Distribution: Uganda.
Interception: ship's hold carrying E. African produce

Opogona sacchari (Bojer) – – sugar cane borer
Distribution: occurs on all the islands surrounding Africa.
Interception: Mauritius: green ginger

Setomorpha tessellatella (Blanchard)
(= *Lindera tessellatella* Blanchard)
Distribution: South America, California, Australia, New Zealand, Fiji, Britain.
Interceptions: Guyana: fish maws
Australia: green ginger

Tinea cloacella Haworth, see *Nemapogon cloacella* (Haworth)
Tinea columbariella Wocke
Distribution: Europe, North America.
Interceptions: Sweden: dried skins
Turkey: cargo unknown
Tinea vastella Zeller, see *Ceratophaga vastella* (Zeller)

Tortricidae

Argyroploce leucotreta Meyrick, see *Cryptophlebia leucotreta* (Meyrick)

Carpocapsa spp., see *Cydia* spp.

Cryptophlebia leucotreta (Meyrick)
(= *Argyroploce leucotreta* Meyrick)
  Distribution: Ethiopian region.
  Interception: Kenya: sunflower seed

Cydia amplana (Hübner)
  Distribution: continental Europe.
  Interception: France: walnuts

Cydia pomonella (L.) – – codling moth
(= *Enarmonia pomonella* (L.))
(= *Laspeyresia pomonella* (L.))
(= *Carpocapsa pomonella* (L.))
  Distribution: worldwide in apple-growing districts.
  Interceptions: France: walnuts
              Italy: walnuts
              Turkey: gallnuts

Cydia splendana (Hübner) – – acorn moth
(= *Enarmonia splendana* Hübner)
(= *Laspeyresia splendana* (Hübner))
  Distribution: north-west Europe.
  Interceptions: Portugal: chestnuts
              Spain: chestnuts

Enarmonia spp. see *Cydia* spp.

Laspeyresia spp., see *Cydia* spp.

Diptera

Nematocera

Bibionidae

Dilophus febrilis (L.)
  Distribution: Europe.
  Interceptions: USA: cottonseed meal,
              Canada: feed barley

Culicidae

Aedes flavescens (Müller)
  Distribution: N. Holarctic.
  Interception: Afghanistan: dried fruit

Anopheles annulipes Walker
  Distribution: Australia, New Guinea.
  Interception: Australia: rice

Culex pipiens L. – – common gnat
  Distribution: north and south temperate zones.
  Interception: Australia: rice

Culiseta annulata (Schrank) – – banded mosquito

(= *Theobaldia annalata* Schrank)
  Distribution: western Palearctic.
  Interception: Holland: feather meal

Theobaldia annulata Schrank, see *Culiseta annulata* (Schrank)

Scatopsidae

Scatopse notata (L.)
  Distribution: cosmopolitan.
  Interceptions: Brazil: Brazil nuts.
              Pakistan: guar meal, rapeseed meal

Trichoceridae

Trichocera regelationis L.
  Distribution: Europe, N. America.
  Interception: empty ship's hold

Stratiomyidae

Hermetia illucens (L.)
  Distribution: North, Central, South America, West Indies, tropical and warm temperate zones of Mediterranean, Europe, Asia, Africa, Australia, New Zealand.
  Interceptions: USA: wet-damaged rice,
              Jamaica: coconut shell

Cyclorrhapha

Calliphoridae

Calliphora augur nociva Hardy
(= *Calliphora nociva* Hardy)
  Distribution: Australia.
  Interception: Australia: hides, sheepskins

Calliphora erythrocephala Meigen, see *Calliphora vicina* (Robineau-Desvoidy)

Calliphora lata Coquillett
  Distribution: Taiwan, Japan, Korea, N. China, Siberia.
  Interception: Japan: tuna fish

Calliphora nociva Hardy, see *Calliphora augur nociva* Hardy

Calliphora stygia Fabricius
  Distribution: Australia, New Zealand.
  Interception: Australia: sheepskins, wheat gluten stowed above meat residues

Calliphora vicina (Robineau-Desvoidy) – – blue bottle
(= *Calliphora erythrocephala* Meigen)
  Distribution: Europe, North and South America, Pakistan, north India, south China, Japan, Korea, Manchuria, Siberia, Egypt, Ethiopia, Australia, Tasmania, New Zealand.

Interceptions: South Africa: fishmeal
      India: jaggery
      Australia: sheepskins

Chrysomya putoria Wiedemann
  Distribution: Africa south of the Sahara.
  Interception: Nigeria: hoof and horn, bones

Chrysomya rufifacies (Macquart)
  Distribution: Australian and Oriental regions.
  Interception: Australia: sheepskins

Lucilia cuprina (Wiedermann)
  Distribution: widely distributed throughout the tropics, sub-tropics and warm-temperate regions.
  Interception: Australia: sheepskins

Lucilia sericata (Meigen) – – sheep maggot fly
  Distribution: world-wide in temperate regions.
  Interceptions: Lebanon: bones
      Japan: in a tin of tuna fish
      Australia: sheepskins, hides

Phormia regina (Meigen)
  Distribution: Holarctic.
  Interception: Canada: feathermeal

Phormia terrae-novae Robineau-Desvoidy – – black bottle
(= *Protophormia terrae-novae* (Robineau-Desvoidy))
  Distribution: north temperate region, sub-Arctic and Arctic.
  Interception: herring meal residues, probably of Scandinavian origin

Pollenia rudis (Fabricius) – – cluster fly
  Distribution: Europe, Asia, North America.
  Interception: Algeria: rapeseed meal

Protophormia terrae-novae (Robineau-Desvoidy) see *Phormia terrae-novae* Robineau-Desvoidy

Ephydridae

Discomyza eritrea Cresson
  Distribution: Port Sudan.
  Interception: Zanzibar: bags of shells

Heleomyzidae

Tephrochlamys rufiventris (Meigen)
  Distribution: Holarctic.
  Interception: France: wheat

Hippoboscidae

Hippobosca camelina Leach
  Distribution: Middle East and North Africa.
  Interception: Eritrea: lentils

Melophagus ovinus (L.) – – sheep ked
  Distribution: worldwide, temperate.
  Interception: USA: sheepskins

Milichiidae

Desmometopa m-nigrum (Zetterstedt)
  Distribution: nearly-cosmopolitan, not in cool temperate regions.
  Interceptions: Nigeria: hooves and horns.
      Manchuria: wet damaged maize.

Fannia canicularis (L.) – – lesser house fly
  Distribution: throughout temperate regions.
  Interceptions: Canada: soyabean meal
      USA: pea beans
      Denmark: fishmeal
      Faroes: fishmeal

Fannia scalaris (Fabricius) – – latrine fly
  Distribution: throughout temperate regions.
  Interception: Canada: feathermeal

Musca domestica (L.) – – house fly
  Distribution: worldwide.
  Interceptions: numerous; probable strays from dockside fauna.
      Sub-species *vicina* was taken from Mauritius ginger and *nebulo* from Nigerian groundnut cake

Muscina stabulans (Fallén) – – false stable fly
  Distribution: cosmopolitan.
  Interceptions: USA: soyabean meal
      France: maize (probable strays from dockside fauna)

Ophyra aenescens Wiedemann
  Distribution: southern USA to Argentina, Galapagos Islands, southern Europe, Hawaii, Marquesa Is., Tahiti, Micronesia.
  Interception: Argentina: hooves and horns

Stomoxys calcitrans (L.) – – stable fly
  Distribution: cosmopolitan.
  Interception: ship's hold

Phoridae

Diploneura cornuta (Bigot), see *Dohrniphora cornuta* (Bigot)

Dohrniphora cornuta (Bigot)
(= *Diploneura cornuta* (Bigot))
  Distribution: cosmopolitan.
  Interception: Brazil: Brazil nuts

Megaselia septentrionalis (Schmitz)
  Distribution: Europe.
  Interception: freight container, probable stray from dockside fauna

Tephritidae
(= **Trypetidae**)

Ceratitis capitata (Wiedemann) – – Mediterranean fruit fly

Distribution: widespread in fruit growing countries with Mediterranean climate.

Interceptions: Greece: currants
Rotten Spanish oranges in store.

Trypetidae, see Tephritidae

Siphonaptera

Ceratophyllus columbae (Gervais)
Distribution: ranges broadly from the British Isles and Spain to the European part of Russia (Lewis, 1975).
Interception: Jamaican coconut shell — there is little doubt that this specimen is a stray from the British fauna.

Ctenocephalides felis (Bouche) – – cat flea
Distribution: Europe.
Interception: Brazilian crushed hooves and horns — a stray on this cargo

Monopsyllus anisus (Rothschild)
Distribution: temperate north-eastern Asia and Japan.
Interception: Japan: used sacks

Nosopsyllus fasciatus (Bosc)
Distribution: more or less cosmopolitan.
Interception: ship's hold

Hymenoptera

Symphyta

Siricidae

Sirex noctilio Fabricius – – steel-blue wood wasp
Distribution: Europe, Canada, New Zealand, Tasmania, Australia.
Interceptions: Finland: softwood
Holland: box of hams
New Zealand: wooden cases of cheese in store

Tremex columba (L.)
Distribution: Nearctic.
Interception: USA: wooden crates containing machinery

Tenthredinidae

Loderus eversmanni (Kirby)
Distribution: central and north Europe to E. Siberia.
Interception: France: maize

Apocrita-Parasitica

Ichneumonidae

Herpestomus arridens (Gravenhorst)

Distribution: Europe.
Interception: Spain: almonds

Stilpnus gagates (Gravenhorst)
Distribution: Europe.
Interception: France: wheat

Braconidae

Blacus humilis (Nees)
(= *Blacus trivialis* Haliday)
Distribution: Europe, N. America, Mongolia.
Interception: France: walnuts

Blacus trivialis Haliday, see *Blacus humilis* (Nees)

Doryctes parvus Muesebeck
Distribution: Puerto Rico, Cuba, India, Malaya and Australia.
Interception: in pulverised wood, with *Dinoderus* sp.

Euscelinus sarawacus Westwood
(= *Sbeitla furax* (Wilkinson))
Distribution: India.
Interceptions: Tanzania: sisal baler twine
India: wooden cases infested with *Sinoxylon* sp.
bombay ducks
Pakistan: dried fruit
Sri Lanka: nutmegs

Monolexis atis Nixon
Distribution: Australia, Japan, Italy
Interceptions: India: cases of cashew nuts, unspecified foodstuffs.
Australia: cases of corned beef, wood infested with *Lyctus* sp.

Orthostigma pumila (Nees)
Distribution: Palaearctic, across Europe, Russia and Mongolia.
Interception: Cyprus: kibbled carobs

Platyspathius pictipennis Viereck
Distribution: Africa, Madagascar.
Interception: Nigeria: cocoa beans, cottonseed

Sbeitla furax Wilkinson, see *Euscelinus sarawacus* Westwood

Chalcidoidae

Chalcididae

Dirhinus himalayanus Westwood
(= *Dirhinoides pachycerus* (Masi))
Distribution: India, S.E. Asia.
Interception: India: groundnut meal

Dirhinoides pachycerus (Masi), see *Dirhinus himalayanus* Westwood

Cynipoidea

Cynipidae

Synergus ruficornis Hartig
  Distribution: Britain, Europe.
  Interception: Turkey: figs

Apocrita-Aculeata

Apoidea

Apidae

Apis mellifera L. – – honey bee
  Distribution: almost worldwide.
  Interceptions: Israel: in a can of grapefruit
    Spain: in a can of fruit

Formicoidea

Formicidae

Camponotus acvapimensis Mayr
  Distribution: West Africa.
  Interception: in the hold of a ship from West Africa
Camponotus compressus (Fabricius)
  Distribution: India, Pakistan, Sri Lanka.
  Interception: Pakistan: bones
Camponotus herculeanus (L.)
  Distribution: North America, Canada, Europe.
  Interceptions: Canada: wheat and flour, machinery
    southern United States: rice bran, timber, wood dunnage
Camponotus lateralis (Olivier)
  Distribution: southern Europe, North Africa.
  Interceptions: Morocco: peas
    Syria: barley
Camponotus maculatus (Fabricius)
  Distribution: tropical
  Interception: East Africa: cottonseed cake and sawn timber
Camponotus truncatus (Spinola)
  Distribution: Mediterranean region.
  Interception: Morocco: peas
Camponotus variegatus (F. Smith)
  Distribution: India, Sri Lanka.
  Interception: ship returning from India
Crematogaster scutellaris (Olivier)
  Distribution: southern Europe, Mediterranean region.
  Interception: Greece: currants
    Morocco: peas, cork
    Spain: almonds and raisins, cork

Dorylus fulvus (Westwood)
  Distribution: Africa.
  Interception: Libya: almond kernels
Hypoponera punctatissima (Roger)
(= Ponera punctatissima (Roger)
  Distribution: cosmopolitan including Britain.
  Interception: on logs from West Africa
Iridomyrmex humilis (Mayr) – – Argentine ant
  Distribution: worldwide.
  In temperate regions it can only survive in heated buildings
  Interceptions: South Africa: cased and tinned fruit
    Greece: dried fruit
    Portugal: almonds
    Spain: raisins
Iridomyrmex rufoniger (Lowne)
  Distribution: Australia.
  Interception: Australia: dried fruit
Lasius brunneus (Latreille) – – brown ant
  Distribution: Palaearctic Europe and Asia.
  Interception: Iran: dried apricots
Monomorium destructor (Jerdon)
  Distribution: almost cosmopolitan.
  Interceptions: United States (California): raisins
    East Arica: bulkhead of ship near Kenyan coffee
    Hong Kong: noodles
    India: cottonseed cake
    Malaya: coconut shell
    Singpore: illipenuts
    Gilbert and Ellis Islands: trunk of personal effects
Monomorium gracillimun (F. Smith)
  Distribution: India, Southern Europe, North Africa.
  Interceptions: India: hold of ship
    Egypt: from a parcel intercepted by customs
Monomorium pharaonis (L.) – – Pharaoh's ant
  Distribution: cosmopolitan, established in temperate parts of the world in heated buildings.
  Interceptions: East Africa: coconuts and coconut shell.
    Nigeria: cocoa butter
    Bangladesh: dried fish
    Hong Kong: green chillies
    India: seeds and spices
    Philippines: desiccated coconut
    Singapore: illipenuts
    Sri Lanka: desiccated coconut
Monomorium subopacum (F. Smith)
  Distribution: drier parts of the Mediterranean region, West Africa.

Interception: West Africa: sawn timber

Oecophylla longinoda (Latreille)

Distribution: Tanzania, Uganda, Ghana, Northern Nigeria.

Interception: a banana store, almost certainly from African produce

Paratrechina longicornis (Latreille)

Distribution: throughout the tropics. In Britain in buildings.

Interceptions: West Indies: coconut shell
Brazil: Brazil nuts
Mozambique: cashew kernels
Nigeria: sheanuts
Sudan: sesame cake
West Africa: coffee
Burma: rice
India: unspecified foodstuffs
Singapore: sago flour
Sri Lanka: desiccated coconut

Pheidole megacephala (Fabricius)

Distribution: throughout the tropics; occasionally in hothouses, bakeries and restaurants in Britain (Wilson, 1942).

Interceptions: Brazil: Brazil nuts
East Africa: cottonseed cake
Kenya: coconuts
Malawi: coconut shell
Nigeria: groundnut cake, sheanuts
Tanzania: alanblackia nuts
West Africa: palm matting
Sri Lanka: desiccated coconut

Polyrhachis hector Smith

Distribution: Oriental and Indo-Australian regions.

Interception: Hong Kong: miscellaneous cargo

Ponera punctatissima Roger, see *Hypoponera punctatissima* (Roger)

Rhytidoponera metallica (F. Smith)

Distribution: Australia.

Interception: Australia: panicum seed

Solenopsis geminata (Fabricius)

Distribution: tropical regions of the Old and New Worlds.

Interception: Thailand: desiccated coconut

Tetramorium bicarinatum (Nylander)

(= *Tetramorium guineense* (Fabricius))

Distribution: worldwide, but in temperate regions it can only survive in heated buildings

Interceptions: Brazil: Brazil nuts
Nigeria: groundnut cake, bones
West Africa: logs
Hong Kong: cotton goods
Sarawak: sago flour

Tetramorium caespitum (L.)

Distribution: Palaearctic and America.

Interceptions: France: walnuts
Turkey: sultanas, dried apricots

Tetramorium guineense (Fabricius), see *Tetramorium bicarinatum* (Nylander)

Tetramorium simillimum (F. Smith)

Distribution: Ethiopian, Oriental and Indo-Australian regions and also the Pacific islands.

Interceptions: East Africa: coconut shell and coconuts
Singapore: rice residues

Sphecoidea

Sphecidae

Chalybion caeruleum (L.), see *Chalybion californicum* (Saussure)

Chalybion californicum (Saussure)

(= *Chalybion caeruleum* (L.))

Distribution: North America.

Interception: United States: one specimen emerged from the mud nest of *Sceliphron caementarium* (Drury) attached to a used vehicle

Sceliphron caementarium (Drury)

Distribution: North America, Fiji, Samoa.

Interception: a mud nest of this species found attached to a used vehicle

Sceliphron spirifex (L.)

Distribution: Africa and the Mediterranean region.

Interception: Nigeria: groundnut cake

Trypoxylon neglectum Kohl, see *Trypoxylon politum* (Say)

Trypoxylon politum (Say)

(= *Trypoxylon neglectum* Kohl)

Distribution: North America.

Interception: United States (Mississippi): emerged from the mud cells of a nest of *Sceliphron caementarium* attached to a used vehicle

Vespoidea

Vespidae

Delta emarginatum (L.)

Distribution: Africa and Arabia.

Interception: Assab: found among wool and hides

Polistes dorsalis hunteri Bequaert

Distribution: south west United States, east Mexico, Jamaica.

Interception: United States: rice

Polistes fuscatus pallipes Lepeletier

Distribution: North America, a predator of lepidopterous larvae.

Interception: United States: pea beans